Surgical Rationales in Functional Reconstructive Surgery of the Upper Extremity

Jörg Bahm

Surgical Rationales in Functional Reconstructive Surgery of the Upper Extremity

Jörg Bahm
Department for Plastic, Hand and Burn Surgery
RWTH Aachen University Hospital
Aachen, Nordrhein-Westfalen, Germany

Invited authors
Oskar Aszmann (Vienna, Austria)
Andreas Gohritz (Basel, Switzerland)
Mirjam Mahler (Ulm, Germany)
Benedikt Schaefer (Aachen, Germany)
Frederic Schuind (Brussels, Belgium)

Mentor
Rolfe Birch (Harwich and London, Great Britain)

Drawings
Martin Langer (Münster, Germany)

ISBN 978-3-031-32007-1 ISBN 978-3-031-32005-7 (eBook)
https://doi.org/10.1007/978-3-031-32005-7

This Springer imprint is published by the registered company Springer Nature Switzerland AG
The registered company address is: Gewerbestrasse 11, 6330 Cham, Switzerland

Preface

Some 25 years ago, I decided to dive into the field of obstetric brachial plexus palsy, because I always had loved neurology and working with children. I applied for a residency in Paris with Alain Gilbert, an internationally renowned specialist in nerve microsurgery, and started my education with reading and assisting the maximum of brachial plexus surgery I could. I also performed cadaver dissections to rediscover the complex anatomy.

At that time, I was certainly unaware what I would discover and learn through the years.

I followed all available congresses in Europe and had the chance wherein Alain Gilbert introduced me to more personal meetings, like the Heerlen or Narakas club meeting.

I started the related reconstructive surgery myself in 1997 and founded a dedicated unit in 2000. Since then, my focus has extended to all aspects of functional reconstructive surgery of the upper limb, both in children and adults, including brachial plexus lesions, arthrogryposis, spasticity, and spinal cord injury. Also, there were lower extremity injuries showing up.

I concentrated on both surgical technique and background knowledge, and the increasing curiosity for the latter set the basement for this book project.

Surgical strategy: Patient examination, counselling, building a therapy plan, and increasing feedback of my surgeries—as I had (and still have) the opportunity to follow all my patients myself (I actually call this the "same same" approach).

Learning by meeting colleagues: Through the years, I met a lot of major specialists in brachial plexus surgery all over the world—in congresses, during hospital visits, and overall through personal discussion (*you get the most valuable detailed information between two scientific sessions of a congress while sharing a coffee or a glass of wine with THE colleague answering and explaining right that little thing which is not written in the reference article*).

Now, 25 years have passed so fast, and it seems to me that time has come to summarize the essential knowledge, to formulate clear guidelines, to set limits within the unlimited field of medical knowledge, and to highlight the cornerstones, which will enable the upcoming young colleagues to continue on the track without the need to rediscover everything from the start.

This book is aimed to summarize one surgeon's skills and views in a clear, sharp, and concise way. It will not repeat basic knowledge (which might be found within the general references); it bears the danger of being utmost personal, thus limited, and perhaps biased. Therefore, the reader is encouraged to go beyond, to comment, to explore more (and to let me know his thinking), and to share both knowledge and controversies.

I tried my very best to achieve something lasting.

I thank Alain Gilbert a lot who has supported me continuously for 26 years now and Rolfe Birch who once again as a real friend updated my non-native English language together with a lot of expert remarks. Rolfe passed away only several weeks ago and I keep him in very respectful memory, also through this book.

I am very grateful to my friend Martin Langer, a fine hand surgeon and brilliant artist, who offered a set of beautiful drawings to specific themes, making the book much more attractive.

Gabriele Dünwald accepted to contribute personal graphic work which introduces each chapter; thank you so much for sharing.

I also thank Daniela Heller from Springer who encouraged and followed this project with enthusiasm, and Smitha Diveshan from Springer Nature who managed all the editorial work.

Aachen, Germany Jörg Bahm
June 2023

Contents

1 Decision-Making in Reconstructive Surgery

J. Bahm, *Surgical Rationales in Functional Reconstructive Surgery of the Upper Extremity*,
https://doi.org/10.1007/978-3-031-32005-7_1

Surgical decision-making is the utmost difficult and individual compromise between patient demands and surgical expertise in a specific socioeconomic field.

Whereas the young and rather unexperienced surgeon looks for cases, and thus might try hard to alleviate recruitment criteria, the older colleague has memorized all pitfalls and complications together with success stories and thus is more balanced but might be sometimes less innovative.

As our health systems become continuously more regulated and constraining, considering economic and law issues, the dual relation between the patient and his doctor risks to become more complicated and even biased.

Functional reconstruction is a probably rarer and more hidden specialty, and patient demands may vary a lot. Examination and counselling are very important, and the personal experience of the surgeon is a major factor, related to building trust, developing manual skill, and getting obsessed with good patient outcome.

In the modern society, the medical relation has evolved from *fathership* to a more equal, information-based *partnership*, where the patient clearly becomes the actor and decider.

Nevertheless, he will decide within a frame his ratio allowed him to build, based on the knowledge we as doctors provided to him, altogether with various information taken from the internet, relatives, and friends. Second or third opinions are more frequent nowadays.

Decision-making in reconstructive surgery should be a team approach, including on the medical side the anesthesiologists, various surgical partners, and physio- and ergotherapists. The patient comes up with his partner, children, or relatives. At the end, a treatment plan only succeeds within a trustful, continuous atmosphere, with stable interactors from the begin to the end, which deals with all aspects of the patient's life, without fulfilling ALL demanded tasks.

The reconstructive surgeon does not deal with an object of potential improvement, but with a personality, who has his lifestyle, job, sports, and demands.

The patient remains an individual, and the surgeon cannot refer to extensive lists of standardized procedures. The surgical technique involves a lot of various tissues and their specific handling; the surgeon often transgresses his own specialty, and then he either should cooperate with reliable specialists or be able to himself extend his manual practice and skills to the procedures typical for the neighbor specialty he needs to know.

Reconstruction of function only ends once the patient is reintegrated in his daily life (the goal of *participation* is met) and experiences stable and measurable improved function—so this task goes beyond surgery and includes rehabilitation. The dedication to that whole spectrum influences all decision-making.

Decision-making is not the art of convincing the patient to accept the surgery I plan for him.

It is a compromise between a lot of positions, the synthesis of various discussions, the endpoint of a listing of pros and cons, as a daily happening event, with low or high complexity.

This decision-making can be as passionate and rich as the surgery itself; it may be the basis of sincere friendship in its best and a nightmare struggle at its worst.

1.1 Analysis of Patient's Story and Expectations (Anamnesis)

The patient frequently describes problems in daily life or asks what is feasible in terms of functional improvement. Sometimes, there is the obvious or hidden wish for total reconstruction (restitutio ad integrum), especially if he considers himself as a victim, who lost function he had before (the accident). He wants us to repair it all, to make him recover from all limitations.

He might come up glorifying our expertise and highlighting his trust, thereby negotiating our god-like ability to do it all. No, we are unable to accomplish a miracle. We will disappoint and deceive the patient—like my esteemed colleague Rainer Winkel from Frankfurt said: we need to deceive ("ent-täuschen" in German, meaning also disclose the "Täuschung," i.e., the deception, the illusion). We need to take the risk to lose him—as we are unable to execute his unrealistic wishes.

Table 1.1 A flowchart for proper anamnesis in reconstructive surgery

Actual complaints
Pain issues
Psychology
History of trauma
Former surgeries, illness, medical treatment
Risk factors: nicotine, alcohol, diabetes, obesity, hypertension
Job
Leisure activities
Obvious and hidden expectation
Previous (conservative) treatment, physio- or ergotherapy contacts

Only with this in mind may we start to analyze his expectations.

Table 1.1 gives a flowchart of domains: needs in job, leisure, pain issues, psychological issues like body image (puberty, young ladies), and self-esteem.

We should also analyze the degree of realistic expectation, including the possible result after a long period of rehabilitation.

1.1.1 Giving Enough Time and Space for the Patient's Description

Some patients describe their functional needs with great precision, taking us into their life, daily activity, needs, and leisure activities. This is the environment where our counseling should (and can) take place. The challenge is not the feasibility of a procedure, but its adaptation to the individual patient demands and its capacity to improve the patient's condition and fulfil his expectations, at least partially. We cannot promise a result, but we can do our surgery in a skilled and secure way to set all reasonable conditions for a good recovery. We are aimed to set a safe and helpful environment and to perform our technique according to the actual technical standards, thereby setting the conditions for the best reachable outcome.

Patients may describe pain issues and sensory disorders with a lot of details and images—these are precious moments. As these symptoms are subjective, there is no objective measurement to enlighten us and the outcome will be judged in reference to the initial description.

Pain quality, intensity, and circumstances triggering its onset, and especially the feature of neuropathic pain, become living due to the patient's description that we need to listen carefully. Not only is the verbalization itself beneficial to the patient, but it is also the starting point of a therapeutic process, where the surgical planning and the conservative treatment evaluation will be further steps in a holistic approach.

Pain issues in the upper limb are often brought forward to claim work incapacity, and they may be stakeholders for other diseases, like depression, overuse, chronic inflammation, or a so-far undiagnosed disease (small fiber neuropathy).

The younger surgeon may fear to lose his time or to get misconducted listening to the flow of information given by the patient (and it is always useful to be an active partner, orienting the patient's story with questions, repetitions of specific symptoms, efforts of synthesis), but with years of experience, one will value this so personal and privileged moment where two humans meet in a specific frame and seek for a solution.

1.2 Basics of Medical and Surgical Behavior

What makes someone a "good" *doctor*? There are certainly thousands of individual answers, coming from the society, colleagues, or patients. It is a very personal but obvious question. I got inspired by a book named "Der gute Arzt" ("the good doctor") written by the German social psychiatrist Klaus Dörner at the end of his clinical career. He deals with all aspects of a doctor's behavior and the relations to patients, family, colleagues, and institutions, and he emphasizes that in our relation with the patient, we need to become the other (alter ego), to enter his being so that we become able to feel how are the symptoms, what are the demands, and which treatment is the best for this one individuum. It is not about our achievement, and it is not about the society, economics, or forensic safety: it is about the individual subject.

Around this, we then may add our knowledge, opinions, and evidence-based strategy, but we always start with the total respect of the other human being.

What about a *surgeon* now? A manual worker, who tries to treat with his hands and who wants to repair, to reconstruct, and to build, using crafts and material and addressing specifically organs, soft tissue, bones, vessels, and nerves. The real pleasure of the surgeon should be the time passed in the operation room, being part of a team, together with the anesthetists and nurses, colleagues, and trainees. These should be the "happy hours," although there is a major need of skills and sense of responsibility and leadership too.

The counterpart is the time spent with patients at consultation or on the ward, and talking before and after the operation, from the setting of an indication until the last visit, when a stable result has been reached.

All the rest are individual or social additives, rendering the basic job elegant or hasty, stressful, or deeply fulfilling. The surgeon is rarely a socially validated person; his work is a hidden task.

Through any individual career, we meet so many different colleagues, patients, and representatives from administration and health insurance. Over the years, our surgical personality grows and gets completed—and we might have the feeling that just when we are in a good mood and constitution, we are near to the end … that is where so many colleagues try to continue their job in a private setting, in working abroad, for humanitarian missions … although the turn of generations is a well-known rule since the beginning, for everyone.

What are the specific elements of reconstructive surgery, and especially reconstruction of function? It might be a way to try to make the trauma unhappened. The rebuilding is a return to the situation before ("restitutio ad integrum")—so rarely possible. Then, it is like a shared mourning process—we accompany the patient on a way of improvement, but knowing more or less the possible outcome, the lack of restitution. We take up the dream and transform it into a new reality, bringing in our technical skill, empathy, and leadership. In long treatment plans, we become part of the family, especially when we treat babies and children.

Today, after more than 20 years of experience with this surgery on babies, I sit in front of the young adults they became while I lived my career. They were babies and now they enter the professional and social life of young adults, and they will leave me behind.

The young adults I treat today wonder if I will follow them for some 5 or 10 years more, asking respectfully how long they will continue to find me in the hospital.

The colleagues I teach will make their own evolution, either taking over the job or moving in adjacent or opposite directions, and I stay with them seeing decisions happen.

The surgical work in the operative theatre is never a single-player game—every week, the team spirit is newly challenged and nothing is guaranteed by itself.

Renowned doctors and surgeons have written down their memories—leaving us with a lot of inspiring ideas. I also come up here with personal basic thoughts—knowing perfectly that they are mandatory, but they have to be built and adapted by everybody in his own way.

1.3 Multi-, Inter-, and Transdisciplinarity

A discipline is defined as a group of experts sharing the same body of knowledge. There are several medical and also surgical and other disciplines involved with peripheral nerve reconstruction, like neuro-, plastic, orthopedic, and hand surgeons; physio- and ergotherapists; neuropathologists; and bioengineers.

The neurosurgeon is considered an expert in surgical treatment of pathologies within the central and peripheral nervous system and has particular knowledge in nerve anatomy, physiology of nerve de- and regeneration, direct nerve repair techniques, and microsurgery.

The orthopedic surgeon is specialized in bone and joint static and dynamic corrections and functional surgery of the lower limb, whereas the plastic surgeon deals with soft tissue and microsurgery and performs muscle and tendon transfers as well as small vessel and nerve microsurgery—hand and peripheral nerve surgery.

These attributions are of course country/continent dependent.

Giving an image to possible interactions between disciplines, a multidisciplinary approach assembles several disciplines in parallel, like several fruits in a basket. The routine interaction is interdisciplinary, like fruit pieces in a fruit salad. A transdisciplinary approach raises totally new issues, like in the creation of a smoothie.

1.3.1 The Concept of Transdisciplinarity

Already in 1970, Jean Piaget stated that "a child is not a small adult."

In 1996, the French-Romanian physicist Basarab Nicolescu published a "manifest" about transdisciplinarity, where he developed it as a strong concept, but open and tolerant, transgressing frontiers between disciplines. He cited quantum physics, where the quantum (M. Planck: discontinuity of energy) may be seen as a particle or wave. There is also the time-space indeterminism (Heisenberg) and thus different levels of reality (complexity: multiplication of disciplines).

Stephane Lupasco, a French-Romanian philosopher, introduced the "**included third**": extending the concept of "A and non-A"—in classic physics—to the addition of a third status (being neither A nor "non-A") in quantum physics [1, 2].

Transdisciplinarity approaches are actually seen in **science**, like physics; in **medicine**, like for general practitioners [3]; in psychiatry-psychoanalysis, e.g., in the treatment of psychopathy [4]; emergency care of polytraumatized [5]; oncology [6]; and geriatrics [7].

It has to be distinguished from **translational** medicine ("*from bench to bed*").

We also find it in **nursing**, overall in palliative care [8], and even in **politics**—like in Bhutan, the concept of "gross national happiness."

Examples in reconstructive brachial plexus surgery are:

1. Think beyond the nerve repair one to one
2. Think beyond the nerve alone (nerve transfers from intercostals onto thoracodorsalis and thoracicus longus nerve to reinnervate major target muscles)
3. Think global: glenohumeral arthrodesis may be a late option to recreate a stable and basically mobile shoulder
4. Think about life quality: neuropathic pain treatment, amputation, …

1.4 Imagery and Electrophysiology

It is important to visualize muscle and nerve morphology and their function.

Both magnetic resonance imaging and ultrasound have made great progress, both by enhanced optic discrimination and medical expertise. Both allow the microanatomic examination of a nerve still in continuity, either pre- or intraoperatively.

Muscle shape, trophic state, and fat infiltration are easily recognized on routine ultrasound.

High-energy MRI (up to 7 Tesla magnetic field intensity in research protocols to analyze tiny anatomic structures, like wrist ligaments) allowed to develop the concept of *MR neurography* [9]. Proximal or distal alterations may be found, completing the observed clinical patterns.

1.4.1 Electrophysiology and the Dialogue with Neurologists

Only few neurologists seem dedicated to nerve trauma and functional reconstructive surgery.

In compression neuropathy, electrophysiologic diagnosis is standardized and contributes to decision-making and follow-up. But in brachial plexus trauma, we only get valuable information out of teams with regular interaction between the surgeon and the examiner [10]. Here, a prognostic value of the preoperative exam can be established.

Intraoperative electrophysiology differs between neurosurgeons and plastic surgeons. Whereas the former is customized to neuromonitoring, including nerve conduction velocity (NCV) measurements, the latter use direct electrical stimulation, only to analyze muscular motor responses.

Both sensory and motor evoked potentials have been used in adult brachial plexus surgery, but without conclusive benefit so that more teams would use it in a regular way.

1.4.2 Nerve Fascicle Torsion, Parsonage-Turner Syndrome, and Neurogenic Thoracic Outlet

Parsonage-Turner syndrome is characterized by a sudden onset of painful shoulder muscle palsy with rapid amyotrophy of the deltoid and/or spinati muscles. Many etiologies are discussed, but recent contributions on ultrasound imaging show evident patterns of fascicular torsion and nerve edema. Nerve swelling might occur in an acute nerve inflammation process (like seen in the acute phase of Borreliosis), which will distend the fascicles and allow rearrangement and potential torsion once the edema regresses and disappears. In those cases where the diagnosis of *fascicular twist* was made, emergent decompressive surgery allowed pain relief and functional recovery.

The role of ultrasound (and MRI) in neurogenic thoracic outlet still has to be established, especially the dynamic character of the examination to highlight potential narrowing by crossing bands, ligaments, muscle parts, or even a costoclavicular conflict in specific upper limb positions (abduction—lateral rotation). The risk of a false-negative contribution to the diagnosis cannot be excluded, and the supraclavicular exploration may remain in some defined cases; the only way to exclude or confirm the hypothesis of nerve compression is by means of an open surgical intraoperative analysis.

1.4.3 Muscle Testing

Electromyography does not offer reliable information concerning strength and other physiological muscle capacities, interesting the reconstructive surgeon, like excursion and recruitment. Therefore, preoperative clinical testing is mandatory, but not always feasible. If the same function is provided by several muscles, one might remain doubtful about the real strength of the planned muscle transfer. Fortunately, any muscle excursion may be a bit enhanced when dissecting the muscle out from its aponeurotic sheath.

One example is the frequent tendon transfer of the FCU muscle for weakness of wrist or finger extension. The tension of the tendon may be palpated while locating it at the pisiform, but ulnar wrist flexion may be achieved by the FCR, PL, or even active finger flexors whose tendons also cross the wrist.

1.5 Basic Research

1.5.1 Tissue Engineering of Nerve and Muscle Cells

By Benedikt Schäfer and Astrid Bülow from Aachen.

As a result of an injury to a motor nerve, there is a clinical loss of function. The patient is no longer able to control target muscles voluntarily. One example is the Erb-Duchenne palsy (injury of root C5/C6) with no active abduction in the arm or elbow flexion [11]. As sequelae of this kind of traction trauma, frequently significant nerve gaps result on the level of the trunci or fasciculi.

Two aspects have to be considered concerning regain of muscular function after peripheral nerve injury. First, the injured nerve must be reconstructed to enable the signal transmission between the brain and the target organ. Second, the target muscle must be reinnervated.

The denervation of the muscle following nerval damage leads to wide molecular and structural modifications. To better understand the impact of such an injury, it is worth first taking a look at the physiological condition at the nerve injury site and the effects on muscular function.

1.5.1.1 Neurobiology of Peripheral Nerve Injury and Regeneration

The peripheral nervous system (PNS) consists of afferent and efferent neurons. Their cell bodies are located in the spinal cord, and axonal fibers innervate the target areas, e.g., sensory fields or muscles. The primary function of the PNS is to enable neural transmission between the central nervous system (CNS) and the target organs transmitting sensory information or controlling motor activity [12]. The endoneurium, a connective tissue layer, surrounds each axon. Together with the basement membrane of the Schwann cells (SCs), the endoneurium forms the endoneural sheath. The fibers are grouped into fascicles, with either sensory or motor fibers. The perineurium surrounds these nerve fascicles. Several fascicles may be clustered within the inner epineurium, while the outer epineurium is the most external layer surrounding a peripheral nerve [13]. The glial cells of the PNS are the so-called Schwann cells. These cells are crucial to sustaining the PNS function and aligned along the axons; they are excitable and can detect neuronal bursts. Detection of these activity signals allows SCs to adjust their physiology and generate adequate feedback responses to sufficiently support and control neuronal function [14]. Schwann cells can be distinguished into myelinating (mSCs) and non-myelinating SCs (nmSCs). mSCs ensheath nerves in a layer of myelin, and the nmSCs form a Remark bundle, wrapping several axons [15]. Conduction velocity depends on the myelination of nerves: unmyelinated axons conduct with a speed of 0.5–10 m/s, whereas myelinated axons can conduct with velocities ten times as fast. The elongated SC membrane is wrapped up to hundreds of times around the axon [16, 17]. The myelin sheath wrapped by one SC encapsulates an axon and forms an internodal region. The remaining gap in the sheath is formed when neighboring SCs extend and wrap the myelin; this is called the node of Ranvier. The node is a tiny section of the axon membrane (~1 mm wide) more freely accessible from the extracellular solution, with a high density of sodium channels [18]. The combination of electric isolation of the nerve through myelin sheaths and formation of nodes of Ranvier increases the conduction velocity by minimizing the sites of loss of ionic transfer along the axon by saltatory conduction [19].

Both types of SCs also release neurotrophic factors, such as nerve growth factor (NGF), brain-derived neurotrophic factor (BDNF), neurotrophin-3 (NT3), ciliary neurotrophic factor (CNF), glial cell-derived neurotrophic factor (GDNF), artemin, and vascular endothelial growth factor (VEGF) [20] [21]. The myelinating SCs express myelin basic protein (MBP), peripheral myelin protein 22 (PMP22), P0, myelin-associated glycoprotein (MAG), and myelin and lymphocyte protein (MAL), which are released by SCs [22].

NGF, BDNF, and NT3 belong to the so-called neurotrophic factors. NGF acts on the growth of sensory and sympathetic neurons in the PNS as it leads to an increased innervation density, neuron cell body size, axonal terminal sprouting, and dendritic growth [23]. BDNF is neuroprotective by the protection of hypoxia [24, 25]. NT3 also provides axon outgrowth [26]. CNF and artemin are secreted after injuries or stress to the PNS and prevent the neurons from degeneration [27–29]. An antioxidant effect was described for GNDF, as it can act as an acceptor of free radical oxidation products [30]. MBP, PM22, and MAG participate in the myelination. They are found in the myelin sheath of oligodendrocytes and SCs [31–33].

The peripheral nerve system has a great regenerative potential. After nerve injury, several processes at the site of injury occur. After the distraction of the proximal and distal nerve stumps, the so-called Wallerian degeneration takes place at the distal nerve stump during the first hours and days post-injury [34]. Schwann cells (SCs), which are the glia of the peripheral nerve system, demyelinate, proliferate, and transdifferentiate into so-called repair SCs [35–37]. The gene expression profile of the SCs is reconfigured by the ubiquitin-proteasome system. Among others, this downregulates myelin-associated genes (myelin transcription factor Egr2, MPB, MAG, and periaxin) as myelin is known to contain inhibitors of neurite outgrowth and impede SC survival (by the neurotrophin receptor p75) or inhibit SC migration [20, 38, 39]. Therefore, the myelin debris has to be removed from the injury site. Macrophages recruited by cytokines such as TNF-α, LIF, IL-1α and -1β, or MCP-1 to the injury site support the removal of myelin debris [40]. Neurotrophic factors secreted by SCs (artemin, BDNF, NT3, NGF, VEGF, GDNF, and pleiotrophin) enhance neuron survival and axon elongation [41]. The axons degenerate, their myelin sheath degrades, and microglia and macrophages stimulate the SCs to proliferate by secreting cytokines such as IL-1, leading to the synthesis of nerve growth factor-β (NGF-β) [42]. The peak of proliferation occurs approximately 4 days after injury. Proliferating SCs are restricted to their basal lamina tubes, where they align to form so-called Büngner's bands, which provide supporting substrate and growth factors for regenerating axons [43]. The bands start nerve regeneration from the injury site to the nerve target areas.

The success of functional nerve regeneration is dependent on several factors: the sprouting fibers have to direct towards the proper target organ: Also, the distal stump must have neuronal contact. Otherwise, the SCs denervate chronically and do not participate in nerve regeneration anymore. The bands of Büngner disappear [44]. Due to the lack of neural trophic factors, degeneration of the target organ, in this case the muscle, occurs as a consequence [45, 46].

1.5.1.2 Neurobiology of Muscle Tissue

Skeletal muscle comprises approximately 40% of the total body weight. One muscle consists of multiple myofibers or myocytes, which are postmiotic, multinucleated cells. Groups of those cells are embedded in connective tissue called the perimysium. Several groups of these bundles are in turn surrounded by a layer of connective tissue called epimysium, forming the muscle [47]. Each myofiber is innervated by a single motor axon, whereas the motor axon can innervate several myofibers. The motor neuron and all innervated myofibers build a motor unit. When the motor neuron is stimulated, the action potential leads to a depolarization of the presynaptic membrane and thereby the opening of voltage-gated calcium channels. The influx of calcium causes the vesicles stored in the nerve terminal to fuse with the presynaptic membrane and release acetylcholine (ACh) in the synaptic cleft. At the postsynaptic membrane, the myofibers express nicotinic acetylcholine receptors (nAChR) that open upon binding of ACh and cause the depolarization of the sarcolemma, which leads again to the opening of voltage-gated calcium channels triggering the release of calcium of the sarcoplasmic reticulum. Calcium then binds to the regulatory protein troponin C on the actin thin myofilament. Thereby, actin can interact with another cytoskeletal protein, myosin. The interaction between myosin and actin finally results in the contraction of the muscle fiber, with the consumption of energy in the form of adenosine triphosphate (ATP). The area where the motor neuron connects with the myofiber is called the motor end plate or neuromuscular junction (NMJ).

During the maturation of a myocyte, already myoblasts express the embryonic subtype of the nAChR prior to fusion into multinucleated myocytes. Each subtype of the nAChR consists of five subunits. The embryonic subtype differs from the adult one regarding one subunit. Both are built by two α-, one β-, and one δ-subunit. The pentamer is then completed by an ε-subunit in case of the adult subtype and a γ-subunit for the embryonic type. The switch between both subtypes during the maturation of the muscle is mandatory for a functioning muscle as, among other reasons, the adult type shows an increased calcium influx and metabolic stability in comparison to the embryonic type. The switch occurs after the fusion of myoblasts into a myocyte when the axons of a motor nerve start to sprout towards the muscle.

The nerve releases neurotrophic factors that influence the expression of AChR. Agrin is one of those neurotrophic factors. These factors lead to a clustering of the nAChR in proximity to the building synaptic contact and promote the expression of the adult ε-subunit in the synapse-associated nuclei. At the same time, the expressions of the γ-subunit and nAChR in extrasynaptic nuclei are suppressed. Additionally, muscle activity inhibits the expression of the embryonic subunit, leading to the fact that in mature muscle, the expression of nAChR almost exclusively takes place at the motor end plate, where the receptors form clusters with a pretzel-like morphology. Here, agrin and other neurotrophic factors also promote stabilization of the nAChR clusters [48, 49]. It was shown that knockout mice lacking the γ-subunit suffer from severe muscle weakness and premature death, which highlights the importance of this process [50]. In case of a peripheral nerve injury with consequent denervation of the muscle, the nAChR also plays a crucial role in the cellular changes that occur. As a final event, denervation induces muscle atrophy and weakness.

On the cellular level, a fall of resting membrane potential, ionic imbalance, and accelerated protein catabolism can be observed [51]. Cisterna et al. were able to support the hypothesis that neuron-derived factors may be responsible for the maintenance of normal muscle function and the prevention of muscle atrophy. They showed that ACh, which is released by motor neurons, inhibits the expression of connexin hemichannels (Cx43 and Cx45) through activation of nAChR, preventing an increased permeability of the sarcolemma that would result in a change of the resting potential as part of the atrophic process. Furthermore, after denervation, the integrity of the AChR clusters is compromised. There are dystrophic changes at the NMJ with an initially increased turnover rate of nAChRs, fragmentations of the receptor clusters, and finally loss of motor end plates [52, 53]. Furthermore, the adult receptor subtype is again replaced by the embryonic type due to the lack of neuron-derived factors and muscular activity [54]. In general, the atrophy of a denervated muscle is characterized by an imbalance of protein synthesis and degradation, with protein degradation increasing and apoptosis being activated. However, the molecular processes underlying the development are still poorly understood [53, 55].

Shortly after denervation, muscle stem cells called satellite cells are activated and enter the cell cycle. Satellite cells are located between the basal lamina and the sarcolemma, the cell membrane of myocytes. Usually, these cells remain in the resting phase of the cell cycle [56]. In response to stress or trauma, the cells are activated, leave their niche, and enter the cell cycle. Here, the satellite cells proliferate asymmetrically to preserve the stem cell population and differentiate into myoblasts, which eventually fuse into multinucleated myocytes and contribute in a limited way to muscle growth or regeneration of the muscle after injury [57, 58]. In the event of denervation, this mechanism is meant to maintain the muscle mass until reinnervation eventually occurs. In the course of time, this results in a decrease of satellite cells and in an exhaustion of the stem cell pool [59]. The tissue is progressively substituted by fat and fibrotic tissue by activation of local fibroblasts [56, 60].

The time course of those events is relatively well examined in rodent models for denervation and muscle atrophy and thereby also the time span in which a regeneration by direct nerve repair is expected [61]. There is barely any data regarding this issue for human skeletal muscle, and the existing data support the assumption that there are distinct differences [62]. Clinical studies showed that reinnervation through direct nerve repair in case of peripheral nerve injuries leads to no significant regain of muscle function in the case of nerve injuries that last longer than 12 months and is therefore only performed in more recent injuries [63].

1.5.1.3 Classification of Nerval Injuries and Clinical Impact

These findings underline the necessity of timely nerve coaptation or reconstruction after nerve injury. It is also crucial to determine the injured nerve components to choose the proper reconstructive procedure. Nerve injuries are classified after the findings of Sir Herbert Seddon and Sunderland, depending on the severity of the damage and the involved structures [64]. Nerve injuries are distinguished between neurapraxia, axonotmesis, and neurotmesis. Neurapraxia is the mildest peripheral nerve injury, following, e.g., ischemia. Only the myelin sheath is damaged, which results in an interruption of the saltatory conduction, leading to a complete or partial loss of sensory or motor function. Full recovery without further treatment usually occurs after several weeks.

Axonotmesis is defined by a discontinuity of the axon while maintaining continuity of the sheath structures. The loss of sensory or motor function can be temporary but also permanent, depending on the injured structures and severity of trauma. Surgical therapy is needed in severe axonotmesis, as directed nerve growth is unlikely. Because severed axons can sprout again in the preserved endoneurial tubes, fascicles, and perineurium, spontaneous nerve regeneration is possible. However, this happens if only the axon is injured. If the endoneurial tubes are additionally destroyed, some of the sprouting nerve fibers may stray into the perineurium; connective tissue reactions occur, which obstruct a full recovery. In these cases, surgical therapy with excision of the defect section and primary nerve suture (coaptation) must be considered. If only the perineurium is preserved, but the axon, endoneurium, and fascicles are disrupted, spontaneous regeneration is unlikely and inefficient due to the lack of guiding structures and the interposition of connective tissue. Neurotmesis involves the complete dissection of all nerve components. In this case, surgical intervention is needed. If possible, the epineurium of the proximal and distal nerve stump is sutured so that directed nerve growth can occur. Nerve injuries with small defect sizes of approximately 1 cm can be primarily coaptated in an atraumatic microsurgical technique using non-resorbable sutures [65]. Tension after coaptation leads to poor results meaning decreases in fiber count, nerve density, and percent nerve tissue [66–68]. For small-caliber nerves, only the epineurium is readapted—when a nerve with a large caliber is injured, such as the sciatic nerve (diameter 1.5 cm), it is also necessary to adapt the perineurium. Depending on the location of the nerve injury along the axon (proximal lesion sites tend to regenerate faster than at distal locations), the nerve regeneration rate is 1–3 mm per day [12]. Depending on the location of the injury, it can take several months to years until the function is entirely restored. Nevertheless, with increasing defect sizes, functional recovery is usually incomplete and proper regeneration is complicated by scar tissue forming and degeneration of the distal nerve stump [69]. Prolonged denervation time may induce damage to the target areas, which can persist despite successful nerve regeneration.

One of the most important factors for successful reinnervation is the status of the motor end plates and the nAChR. Gupta et al. showed in a histological study of muscle biopsies of 18 patients suffering from an injury to the brachial plexus that even 1 year after the injury, morphologically normal (pretzel-like) motor end plates existed. This result shows that the motor end plates persist and retain their architecture much longer than it has been reported in murine and rodent studies [70]. These findings hold the potential for not only a time-dependent decision in favor or against a direct nerve repair but also a decision depending on the status of the motor end plates based on a muscle biopsy. Nevertheless, further investigation, including a proper correlation to the clinical development, is needed.

1.5.1.4 Tissue Engineering

In a clinical setting, peripheral nerve injuries that are dated back more than 12 months are no longer treated by reconstructing the nerve because of the poor outcome. It has been shown that a primary coaptation of dissected nerves leads to better results in regaining of function when performed in the first 24 h after the injury [71]. To improve motor function in the affected extremity, secondary operations are performed. Tendon transfers and transplantation of free-functioning muscle flaps are two fundamental procedures. Both procedures are based on the replacement of denervated muscle by autologous muscle tissue, accepting the loss of function as well as donor-site morbidity, such as sensory disorder and formation of seroma, at the donor site. In order to avoid the comorbidities and complications of such procedures, research in the field of tissue engineering has been going on for many years to create an equivalent tissue substitute.

Tissue engineering describes the effort to replace diseased and injured tissues by producing adequate tissue substitutes. It is an interdisciplinary field that applies the principles of engineering and life science [72]. There have been major accomplishments engineering different tissue types and even organoids, which have been primarily used for drug testing or modeling disease biology [73]. Despite this progress, clinical translation is still limited, especially in regard to skeletal muscle, whereas several tissue-engineered nerve conduits have been tested both in vitro and in vivo and are available on the market. Tissue engineering utilizes three main components in different combinations, namely progenitor cells, scaffolds, and growth factors or other signaling molecules [74].

Scaffolds, as carrier structures, are needed to mimic the extracellular tissue of skeletal muscle and should provide cells with an optimal environment for proliferation and differentiation. Thereby, scaffolds can be used as acellular implants promoting and directing endogenous regeneration of the tissue or as cell-laden constructs. The scaffolds have to meet several requirements. They should be biocompatible and support cellular attachment. Furthermore, scaffolds are not meant to serve as permanent implants and are supposed to be degraded over time and replaced by endogenous tissue, optimally at the same rate. Due to the elasticity of the muscles, an important goal in muscle tissue engineering is to create scaffolds with flexible properties in order to imitate the elasticity of the muscle. For nerve tissue engineering, also inner and outer guiding structures—corresponding to the peri-, epi-, and endoneurium—are favorable, as they enhance directed nerve growth. Both permeability and conductivity are crucial for, e.g., the sprouting of blood vessels and the transmission of signals. Scaffolds have been produced with many different biomaterials, which can generally be divided into two main groups: natural and synthetical biomaterials. Synthetic scaffolds that have been used for both nerve and muscle tissue engineering including poly(ε-caprolactone) (PCL), poly(glycolic acid) (PGA), polylactic acid (PLA, PLLA), copolymer poly-lactic-co-glycolic acid (PLG, PLGA), polyurethane (PU), polyethylene glycol (PEG), polypropylene (PP), and polyamidoamine (PAAs) [75] as biodegradable materials, e.g., chitosan, alginate, agarose, collagen type I, silk fibroin, fibronectin, as well as decellularized native extracellular matrices, have been investigated.

The advantages of natural materials include that they are bioactive, which means that they may contain promyogenic growth factors and mimic the native extracellular matrix more closely. Furthermore, they are biocompatible. On the other side, there are limitations like the batch-to-batch variability, immunogenicity, rapid degradation in vivo, and a lack of precise design control. However, with regard to the latter, additive processes such as 3D printing are able to minimize this limitation.

In regard to synthetic materials, it is possible to adapt the mechanical cues and nanoscale topography very accurately. However, with these materials, there is the challenge of biocompatibility and cell adhesion with a simultaneous lack of bioactivity [76].

To overcome the limitations of both materials, combinations of them are often applied. For example, a scaffold is made from synthetic materials, which is combined with a cell-loaded hydrogel [77].

1.5.1.5 Muscle Tissue Engineering

As progenitor cells, for example, satellite cells are used. Satellite cells are very well suited as muscle precursor cells for the cultivation of muscle tissue. However, a problem that has not yet been solved is their limited availability. Satellite cells can be isolated from muscle tissue by cultivating individual myofibrils or by an enzymatic procedure. In the case of the cultivation of individual myofibrils, only a few but highly pure satellite cell populations are obtained, whereas a large number of cells can be isolated via the enzymatic procedure, which represents a mixed cell population and must be further purified [78].

For this reason, another cell population is also used in skeletal muscle tissue engineering, which is already regularly used for the cultivation of other tissue types, namely mesenchymal stem cells (MSCs). MSCs are multipotent progenitor cells that occur in the stroma of various tissue types and can differentiate into multiple cell lineages of the mesodermal lineage, including adipocytes, osteoblasts, chondrocytes, or myocytes. According to the International Society for Cellular Therapy (ISCT), they are characterized by their ability to differentiate into the various cell types of their mesodermal origin, their plastic adherence under standard culture conditions, and a defined expression pattern of specific markers. MSCs are positive for the surface markers CD73, CD90, and CD150 but negative for CD11b, CD14, CD19, CD34, CD45, and HLA-DR molecules, among others [79–81]. These stem cells can be isolated from various types of tissue, such as bone marrow, umbilical cord tissue, or adipose tissue. In particular, the isolation of adipogenic MSCs is easy and efficient compared to the isolation of skeletal muscle cells. Adipogenic MSCs (ASCs) can be isolated in large numbers from intraoperatively obtained adipose tissue or a lipoaspirate after liposuction. It has been shown that ASCs have myogenic differentiation potential. Thus, they differentiate into the myogenic lineage when co-cultured with satellite cells as well as after cultivation in conditioned satellite cell medium or by biophysical stimuli, such as cyclic stretch [82–84].

Co-cultivation of motoneurons and skeletal muscle cells enables formation of neuromuscular junctions. In a 3D bioprinted silk fibroin co-culture, Dixon et al. showed emergence of precise motoneuron integration into preformed myotubes [85].

As scaffolds for skeletal muscle TE, many different materials have been used, as described before. The use of acellular hydrogels derived, e.g., from decellularized extracellular matrix (dECM) has shown promising results in a clinical trial of 13 patients with VML [86].

The use of dECM as the basis of a scaffold for myogenic TE has come into focus in recent years because dECM conserves biological features, such as proteins, proteoglycans, and cytokines, whose composition cannot be constructed from synthetic materials [87]. Thereby, dECM supports cell differentiation, maturation, and tissue formation. Kim et al., e.g., used decellularized porcine skeletal muscle to create a scaffold for skeletal muscle TE. The dECM was further processed into a hydrogel and loaded with a myoblast cell line (C2C12). The scaffold was fabricated by using 3D printing. A directed myotube formation could be achieved. The rate of differentiation was up to 1.8 times higher in directed dECM scaffolds compared to undirected dECM and a directed gelatin scaffold. It can be concluded that either composition or alignment of scaffolds influences the building of myotubes [88].

In vivo studies show the relevance of presence of a nerve for the development of a neo-generated muscle. Using a neurotized AV loop model in rats, Bitto et al. created an axially vascularized muscle. MSCs and satellite cells were co-cultured in a prevascularized isolation chamber, and a motor nerve was added. After 8 weeks, you could show a myogenic differentiation of the cells [89]. Nevertheless, the formation of neuromuscular junctions, as described above, is currently used mainly to establish in vitro models of the neuromuscular end plate and the ameliorative study of diseases thereof. The development of an in vivo model of muscle vascularization and neurotization is a promising step towards TE of functional skeletal muscle. However, many more studies and years of hard work are needed to reach the goal of a clinically applicable TE of functional muscle tissue.

1.5.1.6 Nerve Reconstruction and Tissue Engineering

Despite decades of research in peripheral nerve tissue engineering, the gold standard to reconstruct a peripheral nerve that cannot be primarily coapted is the autologous nerve transplant (ANT).

The reasons for this may lay in the complexity of the processes occurring at the proximal and distal nerve stump after peripheral nerve injury demanding an equally flexible and protecting surrounding, not allowing ingrowth of scar tissue while permitting growth factors and cytokines to pass the scaffold and blood vessels to grow in. While the pure physical requirements can be matched better and better with the progress of material science, the functionalization of nerve guides and imitation of pharmacokinetic and pharmacodynamic of neural growth factors are still advantageous.

The first ANT (bridging a 2 cm defect of the *N. lingualis* in a canine model) was already performed at the end of the nineteenth century by Phillipeaux and Vulpian [90]. While there usually is a sufficient amount of sensory donor nerves to reconstruct, and also extended nerve defect sizes up to 20 cm (cumulatively) entail the advantage of no foreign body reaction when using autologous material [65, 91], there is still a mismatch between the quality and caliber of the donor nerve and the injured nerve. As already described above, donor-site morbidity, such as loss of sensibility or paresthesia, pain, allodynia, cold sensitivity, and functional impairment, has been described after autologous nerve transplantation [92–94].

Regarding these disadvantages, the development of a tissue-engineered nerve is of high clinical interest. In addition to the qualities of a scaffold already mentioned above, a tissue-engineered nerve should consist of both an inner and outer (tubular) guiding structure—corresponding to the peri-, epi-, and endoneurium, as they enhance directed nerve growth. Both permeability and conductivity are crucial for, e.g., the sprouting of blood vessels and the transmission of signals [95]. Other autologous materials, such as arteries, veins, tendons, or bones, leading to minor donor-site morbidity, have been tested for peripheral nerve reconstruction, all leading to poor results in regaining of function, so that those materials are not favorable for nerve reconstruction.

Also, various nerve scaffolds—both from biodegradable and nonbiodegradable materials—have been tested for peripheral nerve regeneration in the last decades [96]. With hollow tubes only providing a protected space for nerve sprouting, tubes with an inner guiding structure, addition of growth factors or stem cells, and pharmacotherapy, several strategies for peripheral nerve regeneration have been tested [97]. In 1982, a 6 mm peripheral nerve gap could be reconstructed using a silicone scaffold [98]. Nevertheless, nonbiodegradable materials are bioinert and deprive the regrowing nerve from growth factors and sprouting of blood vessels. Furthermore, they can lead to foreign body sensations and, especially when implanted next to joints, lack flexibility. Silicone is impermeable for bigger molecules and can even lead to neuropathy due to chronic compression, so that it has to be removed in a second surgery after nerve sprouting.

Collagen, fibrin, agarose, poly (lactic-co-glycolic-acid) (PLGA), and polylactic acid polylactide (PLA) are other materials tested as simple tubes for peripheral nerve regeneration [99]. Collagen is a protein of the extracellular matrix, which is bioactive and biodegradable, hence not providing a high mechanical stability. PGA is also biodegradable but bioinert, leading to extrusion. PLCL also belongs to the biodegradable materials and provides good mechanical characteristics, while degradation in vivo takes a long time and may lead to forming of fistulae and foreign body reactions.

There are some FDA-approved commercially available nerve guiding tubes such as NeuraGen®, NeuroMatrix™, and Neuroflex™ consisting of bovine collagen type I, Neurotube® consisting of PGA, Neurolac® consisting of PLCL, and SaluTunnel™ consisting of PVA. Other frequently used materials for bridging nerve gaps are acellular nerve grafts, which are also commercially available as allografts (Avance®, AxoGuard®). Nevertheless, tubular structures without an inner guiding used for nerve reconstruction lack promising results for a defect of more than 3 cm. Furthermore, Schwann cell migration is not supported, and there is no extracellular matrix and no trophic or mechanic support for the sprouting axons [100]. This is why the concept of a simple hollow tube is not favorable for peripheral nerve generation.

In another step, several concepts of biomaterial-based nerve conduits have been tested. There is an enormous amount of combination of outer guiding structures as already described and inner guiding elements, addition of cells, or growth factors. The focus here lies on the inner architecture of the scaffold. Besides already mentioned collagen, PLCL, and PGA, the commonly used materials are silk, which has a low immunogenicity, leads to cell adhesion through arginine, and has a high stability, while only degrading slowly in vivo, and chitosan, which is also biodegradable, provides cell adhesion, and may have an antibacterial function, while being fragile. Examples for added growth factors are NGF, GDNF, BDNF, NT-3, or VEGF.

An example for the addition of cells to functionalize the scaffolds is Schwann cells. The addition of SCs to various nerve conduits has been tested and found to enhance nerve regeneration in several studies [101–103]. While the results are promising and can reduce the amount of needed autologous nerve material, the gain of autologous SCs still needs a donor nerve with the resulting donor-site morbidity already described. Furthermore, the proliferation of human autologous SCs in an in vitro cell culture is slow and needs several weeks to produce a sufficient number of SCs depending on the defect size—still being at risk of failure. This also means that the nerve reconstruction cannot take place in the same surgery as the nerve extraction leading to a delayed reconstruction, which is also disadvantageous.

Knowing this, cells with similar characteristics and features with a less severe donor-site morbidity and higher availability seem to be favorable for nerve reconstruction. Stem cells from the adipose tissue can be differentiated into so-called SC-like cells (SCLCs). The harvesting of adipose tissue is simple and does not cause relevant donor-site morbidity. The collected adipose tissue can be further processed, ASCs are isolated from the stromal fraction by enzymatic digestion, and ASCs proliferate rapidly in vitro. ASCs secrete various growth factors and cytokines, such as VEGF, HGF, G-CSF, GM-CSF, Il-6, Il-8, IL11 TNFα, M-CSF, and TNF α, indicating their multiple features for regenerative medicine approaches [104]. Although ASCs and SCLCs have been tested with various nerve conduits, no superiority of ASCs or SCLCs has been shown compared to the autologous nerve transplant [97].

Lately, additive manufacturing has also been used for peripheral nerve regeneration. The big advantage of this concept is the precision, personalizability, and adjustment of the manufactured nerve. There are several techniques for additive manufacturing, such as stereolithography apparatus (SLA) using acrylate and methacrylate, digital light processing (DLP), fused deposition modeling (FDM), or inkjet. The used so-called bioink consists of 3D printable materials such as biodegradable hydrogels (alginate, chitosan, collagen, silk fibroin, or gelatin), synthetic hydrogels (PEG, PAM, PU), or polyester [100, 105]. Also, growth factors and signal molecules can be added to the bioink. For example, with a 3D printed scaffold made of alginate, a 10 mm defect in the sciatic nerve of the rat has been bridged, showing Schwann cell migration after 8–14 days and leading to small myelinated fibers after 2 weeks, which are promising results [106].

Nevertheless, and despite decades of intensive research in peripheral nerve regeneration, the autologous nerve transplant still remains the gold standard for nerve reconstruction.

1.5.2 Living Nerve-Muscle Interaction

The **motor end plate** appears to be the connecting point between two highly specialized entities: the peripheral nerve and the muscle target.

The peripheral nerve bears the motor fibers responsible for the motor activity of the specific muscle, and the different motor fiber types condition the muscle activity profile (fast or slow twitch, tonic, or phasic muscle) (see Table 3.2: principal motor fiber types and their relation to the myofiber).

On the other hand, the real work a muscle provides every day will shape the myofiber composition and so will retroactively influence the mosaic pattern/proportion of subtypes, in order to respond to the task (mainly *phasic*, short and intense action, or *tonic*, sustained but smaller power).

As the basic morphologic link, the motor end plate is a subject of many investigations, but so far unfortunately not in the field of peripheral nerve surgery. It is affected in myasthenia and its morphology changes with age (fragmentation of the pretzel-like shape), and it seems to withstand longer periods of denervation … but it may be the "Achilles tendon" of insufficient reinnervation in many of our clinical situations. Hypothetic reasons are multiple: insufficient neural input, architectural or functional disturbance, lack of satellite Schwann cells, and dysfunction of the postsynaptic pathway.

Interestingly, experimental or clinical *direct neuromuscular neurotization* produces interesting results, visible in surgeries of obstetric brachial plexus lesions, where the upper or middle trunk neuroma sends regenerating cones into the nearby scalenus muscle, and also in selected surgical muscle reanimation when due to a loss of the distal segment of the motor nerve, the nerve input is assumed by grafts put directly into the muscle body, near the "motor point," where they stimulate the formation of new motor end plates and induce clinically relevant motor action.

Electrical stimulation of a muscle by means of external or internal electrodes and a specific current not only counteracts denervation amyotrophy and helps the muscle to survive, waiting for reinnervation, but may also condition various activity patterns depending on the current profile.

These examples of interaction around the neuromuscular junction should stimulate more research in this field, as our clinical reasoning and decision-making are often based on rules described above. A better understanding of this link itself in healthy and pathologic situations may clearly allow us to revise and change some of our current paradigms.

In a recent clinical research paper, Gupta et al. [107] focused on the morphologic changes of human motor end plates (MEP) after prolonged denervation and described the MEP fragmentation and dispersion of acetylcholine receptors. He noted that in the first 10 months, *MEP volume and surface area decreased*, although innervated and structural intact MEP persisted for over 3 years.

He therefore suggested before a nerve transfer surgery to perform a target muscle biopsy to analyze the MEP status.

He also observed the regression from usual pretzel-like shape of MEP to a plaque-like morphology, similar to the early embryonic development state.

In his paper, he discusses possible molecular pathways for MEP changes and advocates agrin and matrix metalloproteinase 3 (MMP-3) involved in the maintenance of normal MEP architecture and PKA/cAMP and G alphai2-Hdac4-myogenin-MuRF1 for sympathetic-dependent regulation of the postsynaptic membrane level of acetylcholine receptors by means of recycling and degradation.

1.6 Experience and Psychology

Decision-making is always a multifactorial process. Whereas the young surgeon is driven by the thrill to perform, the older looks back on drawbacks and complications, and overall on insufficient functional results in terms of range of motion, strength, or integration (neglect).

Moreover, a more integrative look at the patient's or his family's psychology and social environment is mandatory. Is the planned surgery or reconstructive plan the most adapted and promising option for this individual patient? Will the predicted result reach the average of previous results of this same operation?

"Psychology" relates to the type of relation and confidence we were able to establish so far. Does the patient seem reliable, trustable for the provided information, and to be able to follow the postoperative rehabilitation plan? Are his expectations reasonable? I learned to fear those patients who put all their confidence in my skills (you know what is best for me … we give our child in your hands) who might become as disappointed at one later moment, compared to the initial sincere, but dangerous enthusiasm.

Experience with children is very particular, as they grow and come back 10 or 20 years later, reporting on how our treatment affected their individual and social living. I actually have this chance to see children who became young adults and who underwent surgery some 20 years ago while being babies. This is a never-ending rich experience, not comparable with any long-term retrospective review. It sharpens the actual indication and at the same time allows a much more global view. After all, an extremity is only one body part, and the personality is built by so many different elements, mainly the character and self-confidence, helping in mastering daily duties in an ever-changing, progressing world.

Experience builds self-assurance and trust in my surgical capacities, but also limits our beautiful job. Experience comes from repetition of procedures, refinement, simplification, optimizing and speeding up of the procedure, and going faster through tissues right to the critical point. Microsurgical experience grows with a sound and clear initial teaching, repetitive exercises, and regular use—than it is a real pleasure—and a guaranty for good quality surgery and self-satisfaction.

When getting older, the patients get younger, and there might be the benefit of "natural authority" just because you are in the age and position of natural confidence. Easy but dangerous. Easy because some discussions turn shorter, and some decision-making on the patient's side seems easier. Probably, the best age for a surgeon could be between 50 and 80.

I experienced to develop thoughts intraoperatively about what would be the functional consequences of my actual manufacture—e.g., when performing stump coaptation of a mixed nerve or a delicate interfascicular nerve graft with the need of very precise positioning of the graft ends onto the recipient slice. This may be helpful and stimulating, but also embarrassing if this idea invades you like a stressor. Beware of excessive noise or parallel triggering while performing delicate surgical steps in the OR!

How far extends our responsibility, from indication through surgery until the final result? This is reflected back to the initial situation of decision-making (with experience) and of course affects our personal intellectual working out.

1.7 Evidence-Based Practice

May we apply this in individualized reconstructive surgery?

These are the principles of evidence-based medicine (EBM):

Principles:

1. Defining a **level of *evidence***:
   Level I: Evidence obtained from at least one properly designed *randomized controlled trial.*
   Level II-1: Evidence obtained from well-designed controlled trials without randomization.
   Level II-2: Evidence obtained from well-designed *cohort studies or case-control studies*, preferably from more than one center or research group.
   Level II-3: Evidence obtained from multiple time series designs with or without the intervention. Dramatic results in uncontrolled trials might also be regarded as this type of evidence.
   Level III: Opinions of respected authorities, based on clinical experience, descriptive studies, or reports of expert committees.

2. Establishing a **level of *recommendation***:
   Level A: Good scientific evidence suggests that the benefits of the clinical service substantially outweigh the potential risks. Clinicians should discuss the service with eligible patients.
   Level B: At least fair scientific evidence suggests that the benefits of the clinical service outweigh the potential risks. Clinicians should discuss the service with eligible patients.
   Level C: At least fair scientific evidence suggests that there are benefits provided by the clinical service, but the balance

between benefits and risks is too close for making general recommendations. Clinicians need not offer it unless there are individual considerations.

Level D: At least fair scientific evidence suggests that the risks of the clinical service outweigh potential benefits. Clinicians should not routinely offer the service to asymptomatic patients.

Level I: Scientific evidence is lacking, of poor quality, or conflicting, such that the risk versus benefit balance cannot be assessed. Clinicians should help patients understand the uncertainty surrounding the clinical service.

While scrolling through Wikipedia, I found the "opposing" concept of "**science-based medicine**" claiming a base of scientific principles, having undergone examination of plausibility, allowing a fact-based judgement and including a skeptical attitude and inquiry.

Honestly, I do not know the protagonists of this concept and the value of their website titling with the same name (science-based medicine), but I may use their basic principles to understand and discuss my own rationale.

1.8 Discussion

According to the principles of evidence-based medicine (EBM), we may be happy in our community to come up with a level II-2; a randomized controlled study is impossible to be conceived with our patient groups. Nevertheless, we stand on a number of scientific principles and a lot of them are described in this book. Are we bad scientists? Is our surgical work not enough validated? How to improve?

One problem arises from the multiple variables influencing our treatment and the multifactorial diagnosis behind it. It is difficult or impossible to design a clear study protocol.

Randomization is obviously impossible due to ethical concern.

Rare are the colleagues who are happy to collect enough similar cases over years to design a cohort study (looking back: thus retrospective … nearly always).

On the other hand, scoring becomes more common, and in units employing ergotherapists (occupational therapists), regular standardized controls with recording of function parameters and quality of life questionnaires add valuable information—but the surgical community often does not know the scores and the value of their results. In a meeting talking about obstetric palsy, there are already different ways to understand the Mallet score—how could we imagine to use different scoring tests and compare the results in terms of function, activities of daily living, or life quality without conforming to our previous tools (see later the iPluto initiative).

I still believe that retrospective cohort studies on specific surgical techniques and their outcome are of real value and educational help. I have read meta-analyses on nerve transfer surgery, which added no knowledge for my daily work, although they sound well and are labeled better in EBM. A contrario, case reports about a new surgical technique may be of great value, although they are regularly excluded from systematic reviews.

It seems to me that EBM has been conceived for great numbers (like in statistics, where the value of the test is raising with the number of samples) and "simple" problems with few changing factors. Also is surgery different, especially when the technique may not be standardized. The outcome will also be affected by the individual training and revalidation process, motivation, and further integration (participation) in daily living activities, especially in professional settings.

We probably have to accept this melting pot, this mosaic of knowledge, results, and critical analysis—but we also should remain open-minded for further developments.

References

1. Nicolescu B. Manifesto of transdisciplinarity. Albany: State University of New York Press; 2002.
2. Nicolescu B. Stéphane Lupasco et le tiers inclus. Rev Synth. 2005;126:431–41.
3. Lynch JM, Dowrick C, Meredith P, McGregor SLT, van Driel M. Transdisciplinary Generalism: naming

the epistemology and philosophy of the generalist. J Eval Clin Pract. 2021;27(3):638–47. https://doi.org/10.1111/jep.13446. Epub 2020 Sep 16.
4. Chartier J-P. From pluridisciplinarity to transdisciplinarity. Cliniques. 2012;1(3):96–114. https://doi.org/10.3917/clini.003.0096.
5. Lapierre A, Gauvin-Lepage J, Lefebvre H. La collaboration interprofessionnelle lors de la prise en charge d'un polytraumatisé aux urgences : une revue de la littérature [Interprofessional collaboration in the management of a polytrauma at the emergency department : a literature review.]. Rech Soins Infirm. 2017;(129):73–88. French. https://doi.org/10.3917/rsi.129.0073.
6. Fernandes I, Rueff MC, Portela S. Transdisciplinarity in strategic decisions for oncological treatments. Med Law. 2015;34(1):645–59.
7. Sargent L, Slattum P, Brooks M, Gendron T, Mackiewicz M, Diallo A, Waters L, Winship J, Battle K, Ford G, Falls K, Chung J, Zanjani F, Pretzer-Aboff I, Price ET, Prom-Worley E, Parsons P, iCubed Health and Wellness in Aging Transdisciplinary Core. Bringing transdisciplinary aging research from theory to practice. Gerontologist. 2020;22:gnaa214. https://doi.org/10.1093/geront/gnaa214. Epub ahead of print.
8. Pétermann M. Transdisciplinarity: a prerequisite for palliative care practice. Rev Int Soins Palliatifs. 2007;22:19–22. https://doi.org/10.3917/inka.071.0019.
9. Bäumer P, Pham M, Bendzus M. MR neurography. Diagnostic imaging modality for the peripheral nervous system. Akt Neurol. 2014;41:461–8.
10. Bisinella GL, Birch R, Smith SJM. Neurophysiological prediction of outcome in obstetric lesions of the brachial plexus. J Hand Surg Br. 2003;28(2):148–52.
11. Gilbert A, Whitaker I. Obstetrical brachial plexus lesions. J Hand Surg Am. 1991;16:489–91.
12. Menorca RMG, Fussell TS, Elfar JC. Nerve physiology: mechanisms of injury and recovery. Hand Clin. 2013;29:317–30.
13. Krug C, Holzbach T, Giunta R. Periphere nervenverletzungen. Handchir Scan. 2015;4:57–69.
14. Samara C, Poirot O, Domènech-Estévez E, Chrast R. Neuronal activity in the hub of extrasynaptic Schwann cell-axon interactions. Front Cell Neurosci. 2013;7:228.
15. Liu B, et al. Myelin sheath structure and regeneration in peripheral nerve injury repair. Proc Natl Acad Sci U S A. 2019;116:22347.
16. Aguayo AJ, Charron L, Bray GM. Potential of Schwann cells from unmyelinated nerves to produce myelin: a quantitative ultrastructural and radiographic study. J Neurocytol. 1976;5:565–73.
17. Ritchie JM, Rang HP. Extraneuronal saxitoxin binding sites in rabbit myelinated nerve. Proc Natl Acad Sci. 1983;80:2803–7.
18. Baker MD. Electrophysiology of mammalian Schwann cells. Prog Biophys Mol Biol. 2002;78: 83–103.
19. Muzio MR, Cascella M. Histology, Axon. StatPearls; 2020.
20. Nocera G, Jacob C. Mechanisms of Schwann cell plasticity involved in peripheral nerve repair after injury. Cell Mol Life Sci. 2020:1–13. https://doi.org/10.1007/s00018-020-03516-9.
21. Bunge RP. Expanding roles for the Schwann cell: ensheathment, myelination, trophism and regeneration. Curr Opin Neurobiol. 1993;3:805–9.
22. Corfas G, Velardez MO, Ko C-P, Ratner N, Peles E. Mechanisms and roles of Axon-Schwann cell interactions. J Neurosci. 2004;24:9250.
23. Rocco ML, Soligo M, Manni L, Aloe L. Nerve growth factor: early studies and recent clinical trials. Curr Neuropharmacol. 2018;16:1455.
24. Binder DK, Scharfman HE. Brain-derived neurotrophic factor. Growth Factors. 2004;22:123.
25. Chen A, Xiong L-J, Tong Y, Mao M. The neuroprotective roles of BDNF in hypoxic ischemic brain injury. Biomed Rep. 2013;1:167–76.
26. Maisonpierre P, et al. Neurotrophin-3: a neurotrophic factor related to NGF and BDNF. Science. 1990;80(247):1446–51.
27. Sendtner M, Stöckli K, Thoenen H. Synthesis and localization of ciliary neurotrophic factor in the sciatic nerve of the adult rat after lesion and during regeneration. J Cell Biol. 1992;118:139.
28. Sendtner M, Kreutzberg GW, Thoenen H. Ciliary neurotrophic factor prevents the degeneration of motor neurons after axotomy. Nature. 1990;345:440–1.
29. Baloh RH, et al. Artemin, a novel member of the GDNF ligand family, supports peripheral and central neurons and signals through the GFRα3–RET receptor complex. Neuron. 1998;21:1291–302.
30. Mishchenko TA, Mitroshina EV, Shishkina TV, Vedunova MB. Antioxidant properties of glial cell-derived neurotrophic factor (GDNF). Bull Exp Biol Med. 2018;166:293–6.
31. Boggs JM. Myelin basic protein: a multifunctional protein. Cell Mol Life Sci. 2006;63:1945–61.
32. Snipes G, Suter U, Welcher A, Shooter E. Characterization of a novel peripheral nervous system myelin protein (PMP-22/SR13). J Cell Biol. 1992;117:225–38.
33. Quarles RH. Myelin-associated glycoprotein (MAG): past, present and beyond. J Neurochem. 2007;100:1431–48.
34. Waller A. Experiments on the section of the glossopharyngeal and hypoglossal nerves of the frog, and observations of the alterations produced thereby in the structure of their primitive fibres. Edinburgh Med Surg J. 1851;76:369.
35. Jessen KR, Mirsky R. The repair Schwann cell and its function in regenerating nerves. J Physiol. 2016;594:3521.
36. Perry VH, Brown MC, Gordon S. The macrophage response to central and peripheral nerve injury. A possible role for macrophages in regeneration. J Exp Med. 1987;165:1218–23.
37. Chen P, Piao X, Bonaldo P. Role of macrophages in Wallerian degeneration and axonal regeneration after peripheral nerve injury. Acta Neuropathol. 2015;130:605–18.

38. Trapp B, Hauer P, Lemke G. Axonal regulation of myelin protein mRNA levels in actively myelinating Schwann cells. J Neurosci. 1988;8:3515–21.
39. Chaudhry N, et al. Myelin-associated glycoprotein inhibits Schwann cell migration and induces their death. J Neurosci. 2017;37:5885.
40. Rotshenker S. Wallerian degeneration: the innate-immune response to traumatic nerve injury. J Neuroinflammation. 2011;8:109.
41. Beuche W, Friede RL. The role of non-resident cells in Wallerian degeneration. J Neurocytol. 1984;13:767–96.
42. Lindholm D, Heumann R, Meyer M, Thoenen H. Interleukin-1 regulates synthesis of nerve growth factor in non-neuronal cells of rat sciatic nerve. Nature. 1987;330:658–9.
43. Salzer JL, Bunge RP. Studies of Schwann cell proliferation. I. An analysis in tissue culture of proliferation during development, Wallerian degeneration, and direct injury. J Cell Biol. 1980;84: 739–52.
44. Weinberg HJ, Spencer PS. The fate of Schwann cells isolated from axonal contact. J Neurocytol. 1978;7:555–69.
45. Lee DA, Zurawel RH, Windebank AJ. Ciliary neurotrophic factor expression in Schwann cells is induced by axonal contact. J Neurochem. 1995;65:564–8.
46. Jejurikar SS, Marcelo CL, Kuzon WM. Skeletal muscle denervation increases satellite cell susceptibility to apoptosis. Plast Reconstr Surg. 2002;110:160–8.
47. Frontera WR, Ochala J. Skeletal muscle: a brief review of structure and function. Behav Genet. 2015;45:183–95.
48. Kalamida D, et al. Muscle and neuronal nicotinic acetylcholine receptors: structure, function and pathogenicity. FEBS J. 2007;274:3799–845.
49. Grassi F, Fucile S. Calcium influx through muscle nAChR-channels: one route, multiple roles. Neuroscience. 2020;439:117–24.
50. Schwarz H, Giese G, Müller H, Koenen M, Witzemann V. Different functions of fetal and adult AChR subtypes for the formation and maintenance of neuromuscular synapses revealed in ε-subunit-deficient mice. Eur J Neurosci. 2000;12:3107–16.
51. Cisterna BA, et al. Active acetylcholine receptors prevent the atrophy of skeletal muscles and favor reinnervation. Nat Commun. 2020;11:1073. https://doi.org/10.1038/s41467-019-14063-8.
52. Yampolsky P, Pacifici PG, Witzemann V. Differential muscle-driven synaptic remodeling in the neuromuscular junction after denervation. Eur J Neurosci. 2010;31:646–58.
53. Castets P, et al. mTORC1 and PKB/Akt control the muscle response to denervation by regulating autophagy and HDAC4. Nat Commun. 2019;10:1–16.
54. Rudolf R, Straka T. Nicotinic acetylcholine receptor at vertebrate motor endplates: endocytosis, recycling, and degradation. Neurosci Lett. 2019;711:134434.
55. Morano M, et al. Modulation of the Neuregulin 1/ ErbB system after skeletal muscle denervation and reinnervation. Sci Rep. 2018;8:5047.
56. Wong A, Pomerantz JH. The role of muscle stem cells in regeneration and recovery after denervation: a review. Plast Reconstr Surg. 2019;143:779–88.
57. Forcina L, Miano C, Pelosi L, Musarò A. An overview about the biology of skeletal muscle satellite cells. Curr Genomics. 2019;20(1):24–37. https://doi.org/10.2174/1389202920666190116094736.
58. Le Grand F, Rudnicki MA. Skeletal muscle satellite cells and adult myogenesis. Curr Opin Cell Biol. 2007;19:628–33.
59. Viguie CA, Lu DAX, Huang SK, Rengen H, Carlson BM. Quantitative study of the effects of long-term denervation on the extensor digitorum longus muscle of the rat. Anat Rec. 1997;248:346–54.
60. Carlson BM. The biology of long-term denervated skeletal muscle. Eur J Transl Myol. 2014;24: 5–11.
61. Carlson BM, Borisov AB, Dedkov EI, Dow D, Kostrominova TY. The biology and restorative capacity of long-term denervated skeletal muscle. Basic Appl Myol. 2002;12:249–56.
62. Chen L, Huang HW, Gu SH, Xu L, Xu JG. The study of myogenin expression in denervated human skeletal muscles. J Int Med Res. 2011;39:378–87.
63. Ruijs ACJ, Jaquet JB, Kalmijn S, Giele H, Hovius SER. Median and ulnar nerve injuries: a meta-analysis of predictors of motor and sensory recovery after modern microsurgical nerve repair. Plast Reconstr Surg. 2005;116:484–94.
64. Seddon HJ. Three types of nerve injury. Brain. 1943;66:237–88.
65. Dahlin LB. Techniques of peripheral nerve repair. Scand J Surg. 2008;97:310–6.
66. Mackinnon SE. New directions in peripheral nerve surgery. Ann Plast Surg. 1989;22:257–73.
67. Millesi H. Nerve grafting. Clin Plast Surg. 1984;11:115–20.
68. Sunderland IRP, et al. Effect of tension on nerve regeneration in rat sciatic nerve transection model. Ann Plast Surg. 2004;53:382–7.
69. Pfister BJ, et al. Biomedical engineering strategies for peripheral nerve repair: surgical applications, state of the art, and future challenges. Crit Rev Biomed Eng. 2011;39:81–124.
70. Gupta R, et al. Human motor endplate remodeling after traumatic nerve injury. J Neurosurg. 2020a:1–8. https://doi.org/10.3171/2020.8.jns201461.
71. Hart AM, Terenghi G, Wiberg M. Neuronal death after peripheral nerve injury and experimental strategies for neuroprotection. Neurol Res. 2008;30:999–1011.
72. Berthiaume F, Maguire TJ, Yarmush ML. Tissue engineering and regenerative medicine: history, progress, and challenges. Ann Rev Chem Biomol Eng. 2011;2:403–30.
73. Nikolaev M, et al. Homeostatic mini-intestines through scaffold-guided organoid morphogenesis. Nature. 2020;585:574–8.
74. O'Brien FJ. Biomaterials & scaffolds for tissue engineering. Mater Today. 2011;14:88–95.
75. Carnes ME, Pins GD. Skeletal muscle tissue engineering: biomaterials-based strategies for the treat-

ment of volumetric muscle loss. Bioengineering. 2020;7:1–39.
76. Gilbert-Honick J, Grayson W. Vascularized and innervated skeletal muscle tissue engineering. Adv Healthc Mater. 2020;9:e1900626.
77. Schäfer B, et al. Warp-Knitted spacer fabrics: a versatile platform to generate fiber-reinforced hydrogels for 3D tissue engineering. Materials (Basel, Switzerland). 2020;13:3518.
78. Syverud BC, Lee JD, VanDusen KW, Larkin LM. Isolation and purification of satellite cells for skeletal muscle tissue engineering. J Regen Med. 2015;3:117.
79. Robey P. 'Mesenchymal stem cells': fact or fiction, and implications in their therapeutic use. F1000Research. 2017;6:524.
80. Dominici M, et al. Minimal criteria for defining multipotent mesenchymal stromal cells. The International Society for Cellular Therapy position statement. Cytotherapy. 2006;8:315–7.
81. Keating A. Mesenchymal stromal cells. Curr Opin Hematol. 2006;13:419–25.
82. Cai A, et al. Myogenic differentiation of primary myoblasts and mesenchymal stromal cells under serum-free conditions on PCL-collagen I-nanoscaffolds. BMC Biotechnol. 2018;18:75.
83. Bajek A, et al. Human adipose-derived and amniotic fluid-derived stem cells: a preliminary in vitro study comparing myogenic differentiation capability. Med Sci Monit. 2018;24:1733–41.
84. Pantelic MN, Larkin LM. Stem cells for skeletal muscle tissue engineering. Tissue Eng B Rev. 2018;24:373–91.
85. Dixon TA, et al. Bioinspired three-dimensional human neuromuscular junction development in suspended hydrogel arrays. Tissue Eng Part C Methods. 2018;24:346–59.
86. Dziki J, et al. An acellular biologic scaffold treatment for volumetric muscle loss: results of a 13-patient cohort study. NPJ Regen Med. 2016;1:16008.
87. Smoak MM, Mikos AG. Advances in biomaterials for skeletal muscle engineering and obstacles still to overcome. Mater Today Bio. 2020;7:100069.
88. Kim WJ, et al. Efficient myotube formation in 3D bioprinted tissue construct by biochemical and topographical cues. Biomaterials. 2020;230:119632.
89. Bitto FF, et al. Myogenic differentiation of mesenchymal stem cells in a newly developed neurotised AV-loop model. Biomed Res Int. 2013;2013:935046.
90. Dellon ES, Dellon AL. The first nerve graft, Vulpian, and the nineteenth century neural regeneration controversy. J Hand Surg Am. 1993;18:369–72.
91. Socolovsky M, Di Masi G, Battaglia D. Use of long autologous nerve grafts in brachial plexus reconstruction: factors that affect the outcome. Acta Neurochir. 2011;153:2231–40.
92. Ijpma FFA, Nicolai J-PA, Meek MF. Sural nerve donor-site morbidity. Ann Plast Surg. 2006;57:391–5.
93. Tada K, et al. Long-term outcomes of donor site morbidity after sural nerve graft harvesting. J Hand Surg Glob Online. 2020;2:74–6.
94. Hallgren A, Björkman A, Chemnitz A, Dahlin LB. Subjective outcome related to donor site morbidity after sural nerve graft harvesting: a survey in 41 patients. BMC Surg. 2013;13:39.
95. Amani H, Kazerooni H, Hassanpoor H, Akbarzadeh A, Pazoki-Toroudi H. Tailoring synthetic polymeric biomaterials towards nerve tissue engineering: a review. Artif Cells Nanomed Biotechnol. 2019;47:3524–39.
96. Gu X, Ding F, Williams DF. Neural tissue engineering options for peripheral nerve regeneration. Biomaterials. 2014;35:6143–56.
97. Rhode SC, Beier JP, Ruhl T. Adipose tissue stem cells in peripheral nerve regeneration—in vitro and in vivo. J Neurosci Res. 2021;99:545–60.
98. Lundborg G, Gelberman RH, Longo FM, Powell HC, Varon S. In vivo regeneration of cut nerves encased in silicone tubes: growth across a six-millimeter gap. J Neuropathol Exp Neurol. 1982;41:412–22.
99. Pfister LA, Papaloïzos M, Merkle HP, Gander B. Nerve conduits and growth factor delivery in peripheral nerve repair. J Peripher Nerv Syst. 2007;12:65–82.
100. Selim OA, Lakhani S, Midha S, Mosahebi A, Kalaskar DM. Three-dimensional engineered peripheral nerve: toward a new era of patient-specific nerve repair solutions. Tissue Eng Part B Rev. 2021;28(2):295–335. https://doi.org/10.1089/TEN.TEB.2020.0355.
101. Hyung S, et al. Dedifferentiated Schwann cells secrete progranulin that enhances the survival and axon growth of motor neurons. Glia. 2019;67:360–75.
102. Hess JR, et al. Use of cold-preserved allografts seeded with autologous Schwann cells in the treatment of a long-gap peripheral nerve injury. Plast Reconstr Surg. 2007;119:246–59.
103. Strauch B, et al. Autologous Schwann cells drive regeneration through a 6-cm autogenous venous nerve conduit. J Reconstr Microsurg. 2001;17:589–95.
104. Kilroy GE, Foster SJ, Wu X, Ruiz J, Sherwood S, Heifetz A, Ludlow JW, Stricker DM, Potiny S, Green P, Halvorsen YD, Cheatham B, Storms RW, Gimble JM. Cytokine profile of human adipose-derived stem cells: expression of angiogenic, hematopoietic, and pro-inflammatory factors. J Cell Physiol. 2007;212(3):702–9. https://doi.org/10.1002/jcp.21068. PMID: 17477371.
105. Singh D, et al. Additive manufactured biodegradable poly(glycerol sebacate methacrylate) nerve guidance conduits. Acta Biomater. 2018;78:48–63.
106. Hashimoto T, et al. Peripheral nerve regeneration through alginate gel: analysis of early outgrowth and late increase in diameter of regenerating axons. Exp Brain Res. 2002;146:356–68.
107. Gupta R, Chan JP, Uong J, Palispis WA, Wright DJ, Shah SB, Ward SR, Lee TQ, Steward O. Human motor endplate remodelling after traumatic nerve injury. J Neurosurg. 2020b:1–8.

2 Pre- and Postoperative Functional Evaluation in Reconstructive Surgery

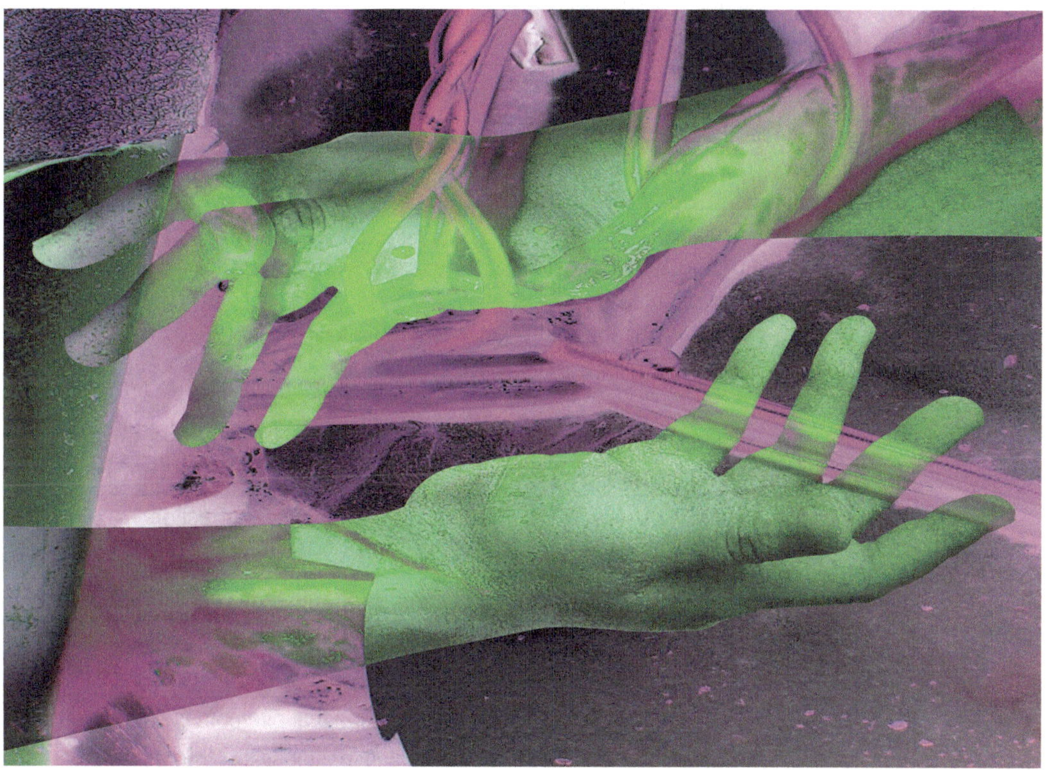

J. Bahm, *Surgical Rationales in Functional Reconstructive Surgery of the Upper Extremity*,
https://doi.org/10.1007/978-3-031-32005-7_2

2.1 *Routine* Tools

Assessment of function regularly means evaluation of motor capacities: motion, antigravity posture, strength, fatigue, and sports capacities.

But in selected situations, it might be recovery of surface sensation or improvement of pain.

2.1.1 Motion and Strength

We regularly assess the passive and active range of motion (ROM) in all involved limb joints, by direct angle measurement (using a goniometer) or visual estimation (e.g., the "quadrant method"; see Fig. 2.1).

A basic information about strength is given through the British Medical Research Council (BMRC) scale from M0 to M5 (Table 2.1), which has been adapted in many ways, e.g., by Gilbert for little children/babies (Table 2.2).

These values do not tell anything about possible fatigue (holding a posture or repetitive movements, eventually holding an object), or the real use in activities of daily living (ADL) or special sport or professional requirements, far beyond any routine or repetitive evaluation.

In a routine setting, we only use the Mallet scale (Table 2.3) or one of its adaptations as a basic scoring system for upper limb motion.

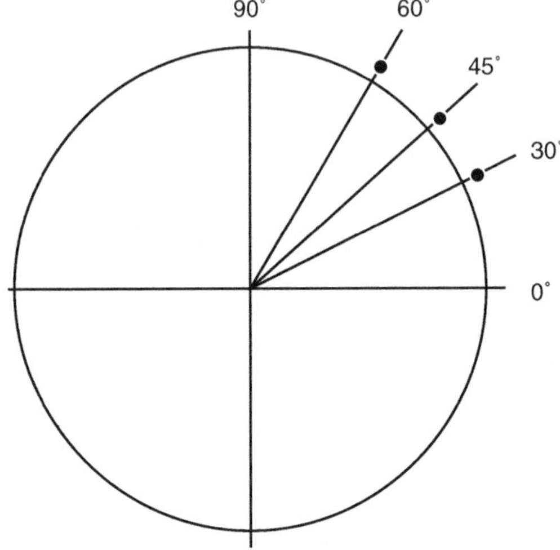

Fig. 2.1 The quadrant method

Table 2.1 BMRC scale

Degree of muscle strength	
M0	No muscle contraction
M1	Muscle contraction not resulting in joint movement
M2	Muscle contraction with movement excluding gravity
M3	Muscle contraction effective against gravity but does not overcome resistance.
M4	Muscle contraction that overcomes some resistance
M5	Normal muscle strength

Table 2.2 Gilbert scale

Stage 0	Complete shoulder flail
Stage I	Abduction or flexion to 45°, no active external rotation
Stage II	Abduction < 90°, external rotation to neutral
Stage III	Abduction = 90°, weak external rotation
Stage IV	Abduction < 120°, incomplete external rotation
Stage V	Abduction > 120°, active external rotation

Evaluation: stage III, IV, V are functionally useful.

Table 2.3 Mallet scale

	II	III	IV
Active abduction	 Inferior to 30°	 30° to 90°	 Superior to 90°
External rotation	 0°	 Inferior ro 20°	 Superior to 20°
Hand to nape of neck	 Impossible	 Difficult	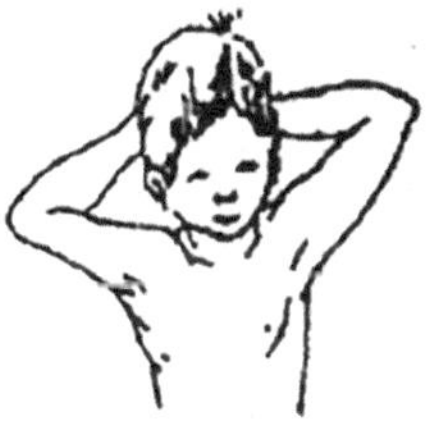 Easy
Hand to back	 Impossible	 S1	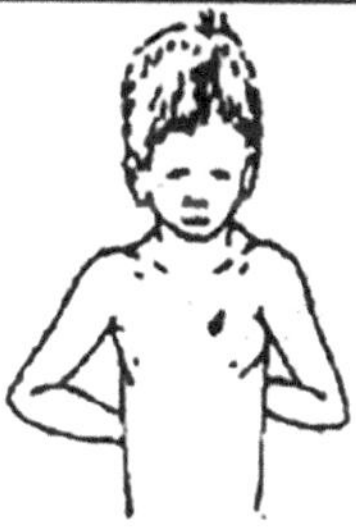 T12
Hand to mouth	 Clarion	 Small clarion	

2.1.2 Sensibility

The slight touch running over the skin allows a first gross evaluation and examination of root-related dermatomes involved (Fig. 2.2).

Moreover, the static or dynamic two-point discrimination test (2PD, Weber) and the use of Semmes-Weinstein monofilaments allow a better investigation of some more specific sensitive modalities and the underlying corpuscles.

Pain is described through reporting of the patient's description or using the visual analogue scale (VAS) from 0 to 10.

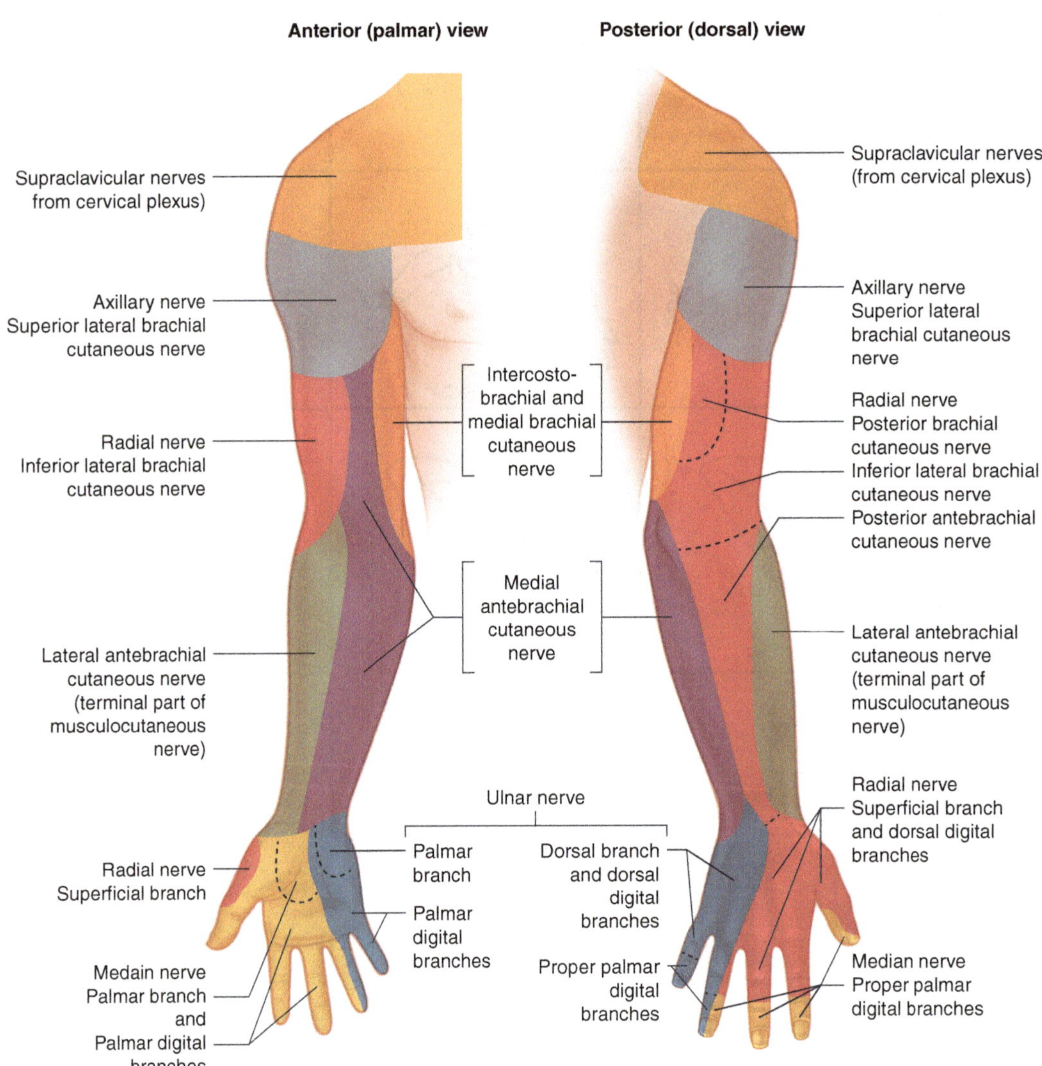

Fig. 2.2 Scheme of skin dermatomes

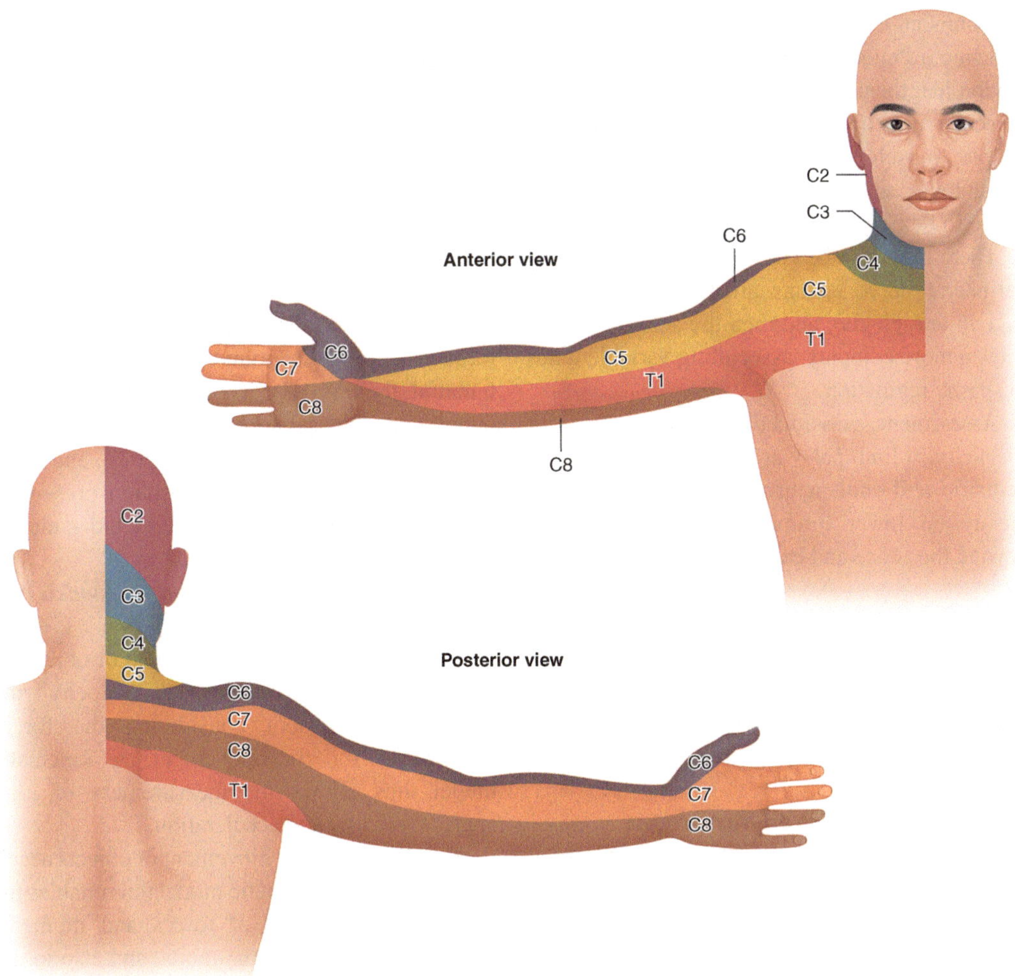

Fig. 2.2 (continued)

2.1.3 Spasticity

Compared to muscle activity after peripheral nerve lesions, the pattern observed in central nervous lesions accompanied by **spastic** muscle activity is more complex and also difficult to be treated.

A central nervous lesion may be associated with a disturbed regulation of the second motoneuron activity, thus leading to involuntary muscle contractions (felt like a hypertonicity) impairing the usual motion pattern, and being velocity dependent, i.e., slow passive counteraction will relief the spastic contraction.

This pathologic neural regulation affects the muscle differently from neurogenic amyotrophy, and there is increasing stiffness related to connective tissue invasion and progressive joint ankylosis (leading to "contractures"). The basic physiological properties of the muscle change, like intrinsic elasticity and contractility (myofiber adaptation), and the tuning between agonist and antagonist are disturbed.

Spasticity and hypertonia must be clearly distinguished from a contracture.

The therapeutic target is either the neural input (Botox, selective partial motor neurotomy, also called hyponeurotization), the muscle itself (detachment of the proximal muscle insertion allowing a "muscle slide" diminishing the tone, tendon lengthening, partial section of the tendomuscular junction), or the agonist-antagonist couple (muscle transfer to augment the more physiologic activity, like wrist extension, preferred to flexion).

There may also be coexistence of central and peripheral nerve lesions affecting the muscles (like in tetraplegia), and then the activity pattern is even more complex, associating variable tonus to spastic regulation.

Assessment, classification, and treatment of spasticity in limb palsy are a specialized field in pediatric and adult neuro-orthopedics, focusing mostly on lower limbs—and going therefore beyond the scope and expertise of our book.

2.2 *Expert* Tools

2.2.1 Electrophysiology (See Sect. 1.4)

2.2.1.1 Video-Assisted Movement Analysis (Vicon): Movement Lab

Since 1996, we have been cooperating on a regular basis with the laboratory for movement analysis within the Institute for Applied Medical Engineering (Helmholtz Institute) at the RWTH Aachen University, Germany (Director: Professor Dr. T. Schmitz-Rode) in the Department of Rehabilitation and Prevention Engineering (Head: Professor Dr. C. Disselhorst-Klug).

Patients with upper limb movement impairment (and especially children with obstetric brachial plexus lesions (OBPL)) were addressed over the years for specific noninvasive electromyography (high spatial resolution (HSR) surface EMG) to study the reinnervation pattern in the biceps muscle. Biceps-triceps co-contractions were then analyzed before and after treatment with botulinum toxin. When the movement analysis was expanded from lower limb gait analysis to the upper limb, we started to apply this technology to record upper limb movements and extended the measurements to various parameters like ROM and key muscle electromyograms.

Since 2005, our experience with the analysis of upper limb movements using the Vicon technology (a motion capturing system made of 10 infrared video cameras and 32 analogue and 16 EMG channels (Vicon 370) to track space and time passive surface markers placed on the upper limbs and the chest of the child) has been applied to children suffering from OBPL complicated by a rotational shoulder imbalance. The aim has been extended to the investigation of forces and torques acting on the joints, especially the glenohumeral joint (GHJ). Therefore, we started to include pre- and postoperative measurements of shoulder forces and torques before and after shoulder release surgery in children presenting different degrees of severity of shoulder medial rotation contracture.

The supporting *biomechanical models* are **kinematic** or **kinetic**.

Kinematics study the movement of objects in space, according to their path, velocity, and acceleration, without considering the forces responsible for the movement. A kinematic model thus allows the measurement of joint angles, velocity, and acceleration.

Kinetics study the movement of objects accelerated by forces. A kinetic model thus enables the additional calculation of forces and moments (torques) applied onto the joints, and also inherent work, power, and energy.

The kinematic model used in our cooperation lab is based on the so-called *rigid segment approach*, where each segment is assigned to one bone or a group of bones (collar bone, scapula, humerus, forearm, hand). Motion between the segments is assumed possible only within the defined joints as a sheer rotation (no translation).

Figure 2.3 shows the segmentation of the kinematic model and the marker configuration, with permanent segment and temporary joint markers. The segment markers form a triad where the marker extremities form a triangle and thus define one plane. Using these triplets allows an assessment of the motion of each segment in all six degrees of freedom. The joint markers (acromion, elbow, wrist) are measured only during a static trial to avoid erroneous data caused by skin movement. The rotational center of the shoulder

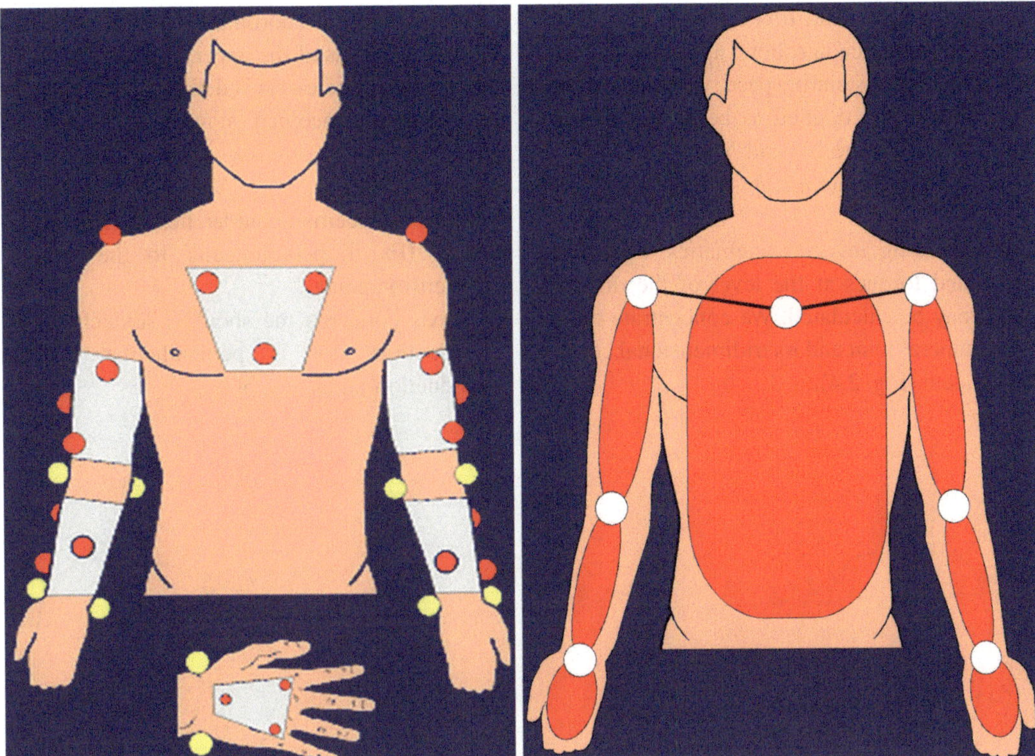

Fig. 2.3 Skin markers/kinematic model

joint has to be estimated as it is located deep under the skin surface. We assume that it lies at a certain distance below the acromion. This distance is assessed individually by a reference measurement for each patient.

The shoulder joint is assumed to work like a ball-socket joint. This model is simple, with some inherent errors. But it allows study of the shoulder motion in daily activities, with or without external load, in a virtual complete ROM. After determination of the joint coordinate system for each joint, the joint motions can be described as relative rotations between these coordinate systems; thus, the kinematic model is completely defined.

The further developed kinetic model to calculate forces and torques is based on the concept of inverse kinematics used in robotics. Inverse kinematics (or inverse dynamics, reverse transformation) applied onto an industrial roboter allows the calculation of joint angles of the robotic arm segments based on the position and orientation of the tool center point, i.e., the end effector. Inverse kinematics is the counterpoint to direct kinematics. The last element of the kinematic chain, the so-called end effector, is moved into the desired position. The other segments must adapt considering the degrees of freedom of their joints. The human arm with its joints also represents a ***kinematic chain***. When we position our hand, the other upper limb joints move into a determined position; these angles may be calculated by inverse kinematics. Difficulties in assessment may arise as there may be different configurations (for a given position of the tool center point) or forbidden ones (mathematically correct but unrealistic for the joint positioning).

The solution of inverse dynamic problems is achieved by algebraic, geometric, or numeric methods. The algebraic method allows the calculation of a homogenous matrix describing the position and orientation of the end effector by a repeated reversal of the Denavit-Hartenberg transformation matrices (a system of conventions, parameters, and transformations defining frames of reference in robotic application, based on a reference system in space, where the Z-axis conventionally is defined as being in line with the joint axis). The X- and Y-axes are defined accordingly to create a right-handed coordinate system.

When using inverse kinematics algorithms, forces and torques at the level of the different joints may be calculated. We concentrate on the glenohumeral joint and its tridimensional coordinate system (Fig. 2.4).

Tridimensional coordinate system at the different joint levels of the upper limb.

Each coordinate axis is defined as the rotation axis for the concerned movement, so for the shoulder:

- X-axis concerns shoulder flexion and extension (Fx is perpendicular to the flexion-extension plane)
- Y-axis concerns the shoulder abduction and adduction (Fy is perpendicular to the abduction-adduction plane)

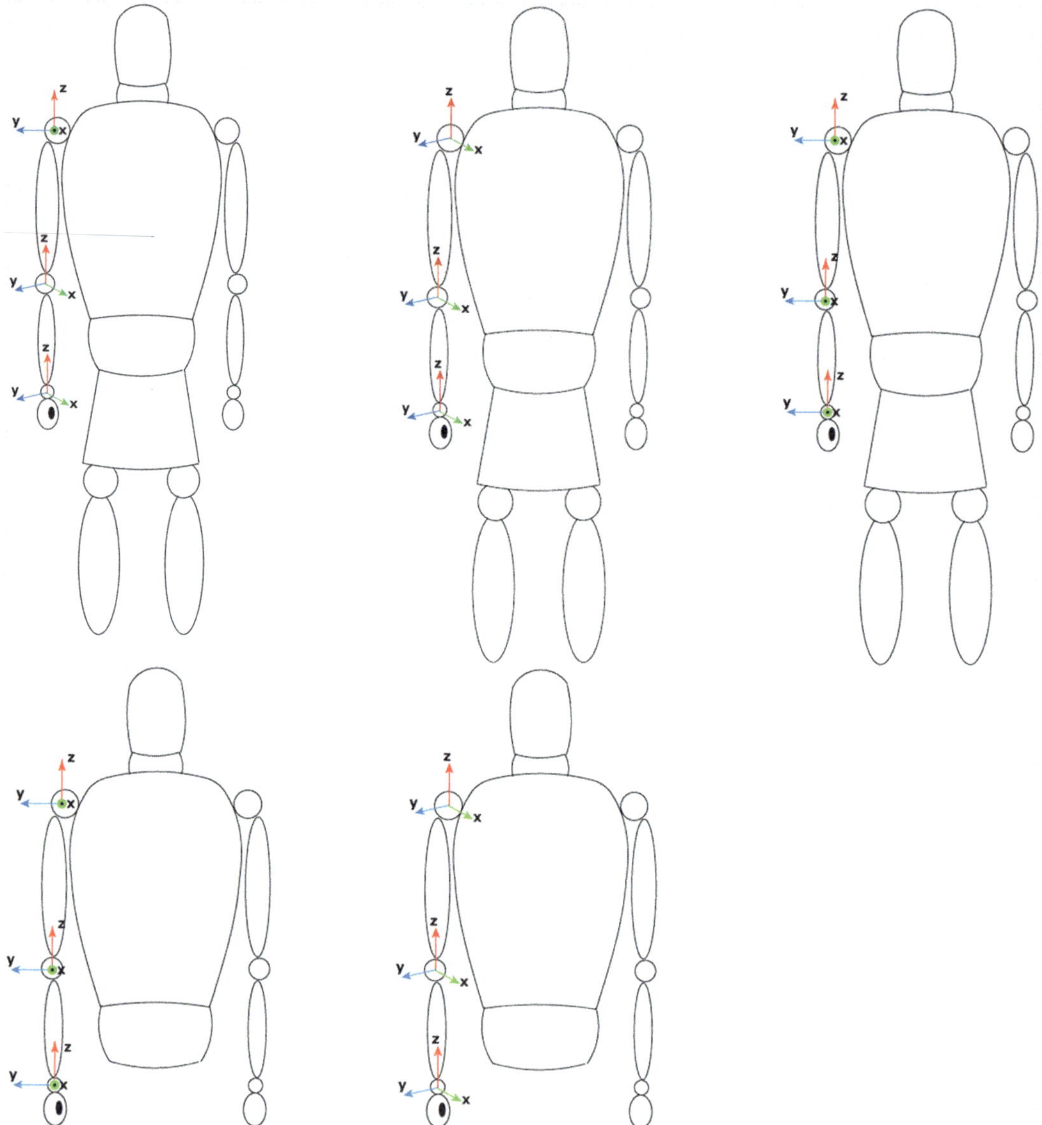

Fig. 2.4 The glenohumeral joint, three dimensional coordinates

- Z-axis concerns the lateral and medial shoulder rotation, identical with the long bone axis, here the humerus (Fz is perpendicular to the rotational plane)

Without any external force, the forces applying in the different directions Fx, Fy, and Fz now may be calculated. If an external force is applied (like when lifting a charge), a force transducer is necessary to integrate these additional forces into the determination of Fx, Fy, and Fz. In our patient series with MRC, we studied both. The force transducer converts measured forces into output signals, further sent to the computers.

The concerned forces mainly correspond to the deforming pressures onto the joint surfaces. Muscle strength in the examined extremity muscles (which all act through a rotational movement around a joint center) is much more represented by the measured torque or lever arm (force * lever arm = torque). Applying these concepts to our MRC patients, and concerning the glenohumeral joint deformations, we were interested in uncompensated or excessive negative Fx (compression onto the glenoid) and negative Fy forces (retropulsion of the humeral head) in the resting arm (adducted along the chest) and negative Fx and positive Fz in the abducted arm (rotation of the coordinate axes). One has to consider that the axes change through any upper limb motion, as any new position of the limb changes the predefined axes of motion direction. When applying the coordinate system to the right or left shoulder, one has to consider that only Fx changes its positive/negative value from right (basic definition) to left; the two other orientations (Fy and Fz) are the same.

2.2.1.2 Motion Patterns (Fig. 2.5)

The child is asked to perform a standardized movement (we initially concentrated on a movement from a position of full medial rotation of the shoulder with an adducted arm to a full lateral rotation with an abducted arm). As not all motion patterns are recorded entirely by the Vicon® system due to marker superposition, some segmented simple movements are necessary to develop a routine examination algorithm (like a figure of eight, straight line, spiral movement, rotation in the YZ plane). To ensure a reproducible movement, the present examination is always performed with a robotic arm. Both the movements of the arm and that of the markers in space are filmed. As the cameras are placed in the examination hall at standardized locations and film the arm under different observer angles,

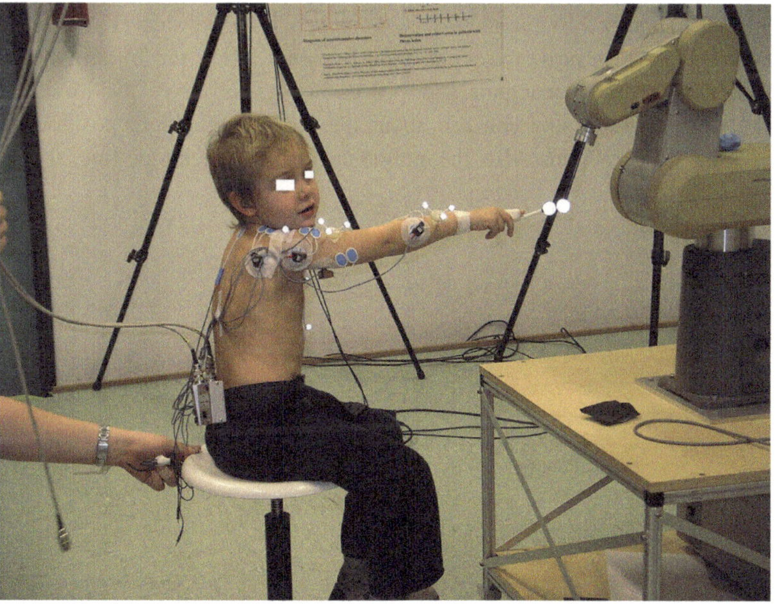

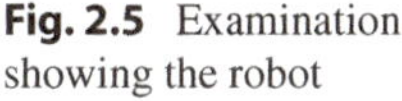

Fig. 2.5 Examination showing the robot

combination of the data allows three-dimensional reconstruction.

Specific software allows to observe the filmed motion sequence and the movement of the markers in space and to calculate time-related curves of the different movements (for the shoulder: abduction and adduction, flexion and extension, lateral and medial rotation). Depending on the study subject, the movement (ROM) analysis is further completed by concomitant recording of the electromyogram (EMG) of selected target muscles, using surface patch electrodes on standardized positions.

External forces are measured using a handle force transducer and calculated through inverse dynamics transformation matrices (see above). Results are presented as time-related curves showing ROM, muscle activity, forces, and torques. Children with various severities of MRC of the shoulder present with different rest positions, as the neutral rotation position of the humeral head is progressively changed into a more medially rotated position. As this affects the correct marker position and labeling procedure before any motion analysis, a corrective "offset" factor corresponding to the severity of MRC had to be added in the recent MRC study group.

There were numerous technical considerations and refinements. It was obviously necessary to adapt the robot to variable sizes of examined children and also to improve the child's security while interacting with a machine and processing a movement without any concern about the individual child's reactions. Also, the motion path had to be adapted to the anthropometric conditions of the examined children (a small child could not cope with very big movements; and a tall child not with very small movements). Also, to estimate the forces, the complexity of the motion pattern had to be simplified.

2.2.1.3 Modeling

As a mathematical simulation, we calculated the forces arising in a standardized shoulder flexion movement under different conditions of humeral head position with increasing medial rotation. Inverse dynamics were applied to an ideal motion dataset and the resulting forces and torques calculated.

Figure 2.6 shows an example of standardized and repetitive movement of the shoulder, as a result of a single-plane flexion-extension tracking exercise following the path of the robotic arm. The motion pattern of the healthy contralateral arm allows the comparison with the observation of reduced joint amplitudes and global alteration of the pathway.

The child shown here has a left OBPP and a severe MRC deformity of the shoulder.

The picture shows the positioning of the passive reflective markers and the EMG surface electrodes. The graph shows the ROM, forces, and torques on both the affected and healthy sides, according to the three motion axes of flexion-extension (Fx, green line), abduction-adduction (Fy, red line), and rotation (Fz, blue line). One observes the limitation of the rotation and the pathologic position of the MRC humeral head (thick blue line, above left) and the major rise of

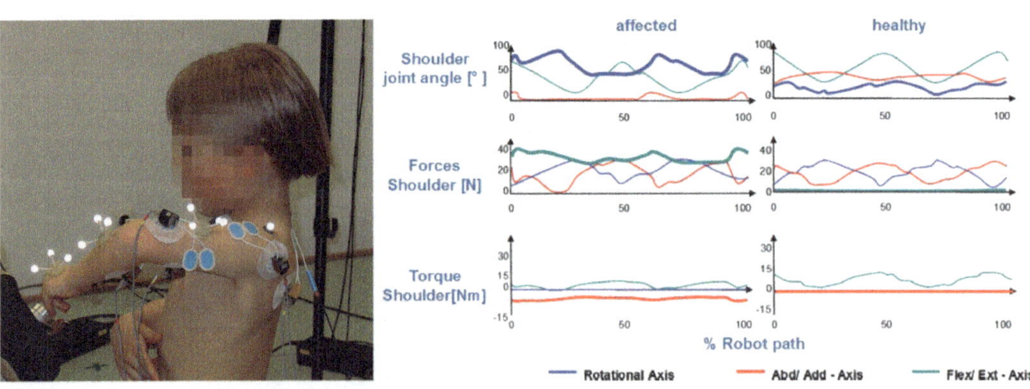

Fig. 2.6 Example of shoulder motion

force along the flexion-extension axis (Fx), measured positive on the affected left side, thus representing a force vector into the glenoid cavity with potential deforming effect. In this situation, Fy and Fz are comparable on both sides.

2.2.2 Further Development of Scoring

Life quality and brachial plexus injuries.

In **adults**, there exist only few investigations, published in English [1] or German [2]. Occupational therapists published in English [3, 4] due to their participation in "brachial plexus programs."

There is a general use of Mallet score [5] or Toronto AMS (active movement scale) [6].

Patient-related outcome measurement scales (PROMS) are done by the patients themselves and are reviewed by the internet-based iPluto (iPluto: international plexus outcome study group) (CHEQ, PODCI, HUH, PROMIS-UE, BPOM, DASH, ABILHAND, PEM-CY, MHQ and WOSI; see www.ipluto.org).

In children, based on the WHO classification ICF (International Classification of Functioning, Disability and Health; see [7]), which analyzes life quality in different dimensions, the Leiden brachial plexus group started in 2013 with a systematic questionnaire and already validated scoring systems (ICF cat, EQ-5DY, WHO QoL, SF-36, DASH, and SCQ) and tried to create OBPP core sets [8, 9].

Since 2016, "iPluto rounds" were organized around specialized centers in the world, and then first results were published [10] showing a provisional ranking dependent on age groups (cutoff 10 years):

Under 10 years: CHEQ, PODCI, HUH, and BPOM

Above 10 years: BPOM, PROMIS-UE, and DASH

A PubMed study done in June 2020 showed that only few studies in children are focusing on "Selbstbestimmtheit" (self-governance, [11]), and there is an obvious lack of concern for such important markers that are functionality, sensation, appearance, compensation mechanisms and limb preference (dominance), self-confidence, body scheme, and explanation will for third persons. We have to bear in mind that we are investigating a very large biopsychosocial field [12].

We actually focus on BPOM [3], the NeuroQoL questionnaire of PROMIS (Patient-Reported Outcomes Measurement Information System) (www.healthmeasures.net), and together with our neuropediatricians, we use CHEQ and DISABkids (in German).

The BPOM tasks are the following:

1. Comb the back head.
2. Bimanual pull grip.
3. Open a large box.
4. Hold a plastic plate with the supinated hand.
5. Hand to mouth (eating).
6. Pass a filament through a needle eye.
7. Drumming.
8. Position a big box above head.
9. Use the PC mouse.
10. Open a bottom at the level of navel.
11. Put the hand in an ipsilateral rear pocket.

2.3 Further Directions

2.3.1 Motion Analysis

2.3.1.1 Pre- and Postoperative Multifactor Movement Analysis

In the future, movement analysis should become a noninvasive, easy-to-perform, repetitive and objective measuring tool to assess the functional status of a limb for diagnosis, pre- and postoperative (comparative) assessment, and also evaluation of physiotherapy. Some prognostic factors might be identified. One of the major problems is the different languages spoken by clinicians and engineers. In my personal experience with regular cooperation with biomedical engineers over 15 years, expectations on both sides are not easily reached. Clinicians tend to consider the technical tools as being magic, easily responding to their pathophysiological and biomechanical questions. Concerning our investigation about forces applied to the glenoid, we expected a software showing force vectors through a defined motion. We have described here the actual common pathway of thinking, which is much more complicated than what we believed initially. Although we got some information about the forces applied to one joint in one movement, along predefined coordinate axes, we are unable to give a generalized view stating that a permanent force vector would be responsible over time for the glenohumeral joint deformation. On the other hand, we learned how a movement in space can be divided in motion vectors along the three axes. We thereby understand better how a global movement within activities of daily life would contain more or less active lateral shoulder rotation, on the affected and healthy sides. We could follow this pattern in a longer motion sequence and analyze all collected data remote to the patient, even by independent objective observers. Our engineers must be constantly connected to our clinical reality; we have to teach them which tools we need for which examinations. Bilateral conferences must become more regular. Motion analysis in the future will be much more than the initial video recording, altogether with some parameters like angles, EMG signals, forces, or torques. It should be a multifactorial tool, with an inherent database of normal datasets (reference), with a memory containing individual prior recordings and a clear examination filter, to answer the specified question. Moreover, this tool must exit research labs to become a natural evaluation tool in orthopedic or physiotherapeutic consultations, being easy to handle, favoring repetitive recordings. Finally, the analysis should be paid for by the insurances, as an objective evaluation tool. We have to admit that so far we collected patchwork, certainly of interest, but far from a routine use to plan our surgeries or to evaluate our conservative or operative treatment. The dream of "modeling" our surgery and study the benefit on a computer, before even doing one incision, will remain such for some further years. But we learned a lot from this exchange with biomedical engineers and we got inspired to analyze movements differently and to understand the real need for biomechanical aspects in functional reconstructive surgery of the upper limb.

2.3.1.2 Further Research: Accelerometers

3D movement analysis is based on 3D tracking systems and allows the objective analysis of upper extremity movements. However, this method is complex and expensive, and translation into clinical practice often fails. Nowadays, accelerometers are used increasingly frequent to determine movement performance. The problem herewith is that no information about the sensor's position can be derived from the accelerometer signals directly. This is why further processing is needed to derive meaningful information about the movement performance from the sensor signals. Based on this context, we work on a new approach based on the use of accelerometers in combination with a classification procedure (like the Mallet score), which allows the quantitative assessment of upper extremity movement performance of patients with OBPP. The methodology should provide direct feedback about the quality of the movement to patients, physicians, and therapists and enable individualized treatment. Three triaxial accelerometers (BMA180, Bosch) are positioned at the manubrium of the sternum, on the distal part of the upper arm, and on the wrist between the distal end of ulna and radius. The sensors are attached directly to the skin of the subjects using a double-sided tape. However, analysis to be applied to the signals has to cope with low deviation in position. Cables are arranged in such a way that they do not hamper the movement of the patient.

2.3.2 Video Analysis, Serious Games, Virtual Reality

With the introduction of the mirror therapy by McCabe, new varieties of cortical learning and adaptative plasticity became suddenly possible. The view of a mirrored healthy hand, making the brain believe that the affected hand is so much improved, has a clear antalgic effect [13].

It took a short time to extend this approach using all digital video technology, allowing new optic or tactile inputs to the patient's brain.

Derived from usual computer video gaming, the "serious" games were born, with this new intention of not to make fun, but to treat a "serious" concern.

Mirror therapy met virtual reality and on the TV screen appeared (like in the mirror) the impaired or even absent hand like a healthy organ, giving the brain the illusion of bimanual activity, space organization, and healthy visual input.

This technology is of course limited to research laboratories and those therapists close to those institutions, but it allows a new and promising insight into our cortical adaptation and learning processes. Especially treatment of severe and chronic complex regional pain syndromes (CRPS) and study of a lot of cerebral sensorial activities show that neuropathic pain affects the brain much more than we believed before (other sensorial modalities may be involved, like hemianopsia in the same area than the concerned limb and its hemicorpus) and the ongoing research will certainly affect our strategies for difficult issues like severe CRPS, limb amputation, and revalidation after severe peripheral nerve injuries.

2.3.2.1 Augmented Reality

My physiotherapy (PT) colleague and friend from Brussels Dominique Mouraux uses 3D augmented reality to improve mirror therapy. This technique consists of adding a virtual image (of the amputated or painful limb) onto the physical world using a computer screen with visual, auditive, and kinesthetic feedback, including motion-detecting video cameras known from video games ("serious gaming"). The use of this enhanced mirror therapy consisting of an interactive video-supported game with motor tasks allows decrease of neuropathic pain [14] by increased immersion. The mechanism behind is still hypothetic and could be a cortical reorganization in primary sensory and motor areas (applied in phantom limb pain and CRPS type II), an increased and focused attention and thus diminishing anxiety, or an increased focus on the limb allowing a reorganized feedback. Functional MRI studies are planned to support these assumptions.

References

1. Rasulić L, Savić A, Živković B, Vitošević F, Mićović M, Baščarević V, Puzović V, Novaković N, Lepić M, Samardžić M, Mandić-Rajčević S. Outcome after brachial plexus injury surgery and impact on quality of life. Acta Neurochir. 2017;159(7):1257–64.
2. Kretschmer T, Ihle S, Antoniadis G, Seidel JA, Heinen C, Börm W, Richter HP, König R. Patient satisfaction and disability after brachial plexus surgery. Neurosurgery. 2009;65(4 Suppl):A189–96.
3. Ho ES, Curtis CG, Clarke HM. The brachial plexus outcome measure: development, internal consistency, and construct validity. J Hand Ther. 2012;25:406–16.
4. Krumlinde Sundholm L, Holmefur M, Kottorp A, et al. The Assisting Hand Assessment: current- evidence of validity, reliability, and responsiveness to change. Dev Med Child Neurol. 2007;49:259–64.
5. Mallet J. Obstetrical paralysis of the brachial plexus. II. Therapeutics. Treatment of sequelae. Priority for the treatment of the shoulder. Method for the expression of results. [in French]. Rev Chir Orthop Reparatrice Appar Mot. 1972;58:166–8.
6. Curtis C, Stephens D, Clarke HM, Andrews D. The active movement scale: an evaluative tool for infants with obstetrical brachial plexus palsy. J Hand Surg Am. 2002;27:470–8.
7. Deutsches Institut für Medizinische Dokumentation und Information (DIMDI). Internationale Klassifikation der Funktionsfähigkeit, Behinderung und Gesundheit (ICF). WHO; 2005.
8. Duijnisveld BJ, Saraç C, Malessy MJA, Brachial TICF, Board PA, Vliet Vlieland TPM, Nelissen RGHH. Developing core sets for patients with obstetric brachial plexus injury based on the International Classification of Functioning, Disability and Health. Bone Joint Res. 2013;2:116–21.
9. Sarac C, Duijnisveld BJ, van der Weide A, et al. Outcome measures used in clinical studies on neonatal brachial plexus palsy: a systematic literature review using the International Classification of Functioning, Disability and Health. J Pediatr Rehabil Med. 2015;8:167–85.
10. Pondaag W, Malessy MJA. Outcome assessment for brachial plexus birth injury. results from the iPluto World-Wide Consensus Survey. J Orthop Res. 2018;36(9):2533–41.
11. Bergman D, Rasmussen L, Chang KW, Yang LJ, Nelson VS. Assessment of self-determination in adolescents with neonatal brachial plexus palsy. PM R. 2018;10(1):64–71.
12. Chang KW, Austin A, Yeaman J, Phillips L, Kratz A, Yang LJ, Carlozzi NE. Health-related quality of life components in children with neonatal brachial plexus palsy: a qualitative study. PM R. 2017;9(4):383–91.
13. McCabe C. Mirror visual feedback therapy. A practical approach. J Hand Ther. 2011;24(2):170–8.
14. Mouraux D, Brassinne E, Sobczak S, Nonclercq A, Warzée N, Sizer PS, Tuna T, Penelle B. 3D augmented reality mirror visual feedback therapy applied to the treatment of persistent, unilateral upper extremity neuropathic pain: a preliminary study. J Man Manip Ther. 2017;25(3):137–43.

3 Biomechanics and Physiology in Reconstructive Surgery of the Upper Limb

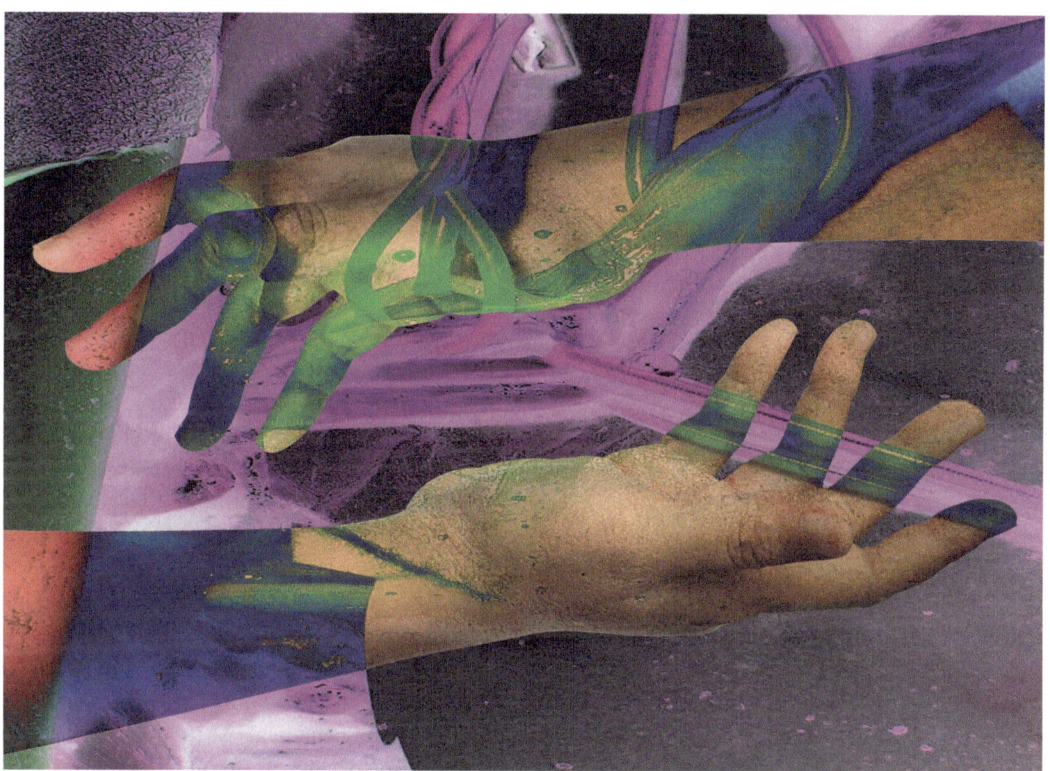

J. Bahm, *Surgical Rationales in Functional Reconstructive Surgery of the Upper Extremity*,
https://doi.org/10.1007/978-3-031-32005-7_3

Table 3.1 Agonists and antagonists in upper limb joint motion

Shoulder
Abduction-adduction: deltoid and supraspinatus vs. pectoralis major
External-internal rotation: infraspinatus and teres minor vs. subscapularis (and eventually dorsi/teres major)
Forward-backward flexion (ante-retropulsion): pectoralis major and anterior deltoid vs. posterior deltoid
Elbow
Flexion-extension: biceps and brachialis vs. triceps
Forearm
Prosupination: pronator teres and quadratus vs. supinator and biceps
Wrist
Flexion-extension: wrist and finger flexors vs. wrist and finger extensors
Radial-ulnar deviation: ECR and FCR vs. ECU and FCU
Fingers
Flexion-extension: flexor superficialis and profundus vs. extrinsic and intrinsic extensors
Abduction-adduction: interossei
Thumb
Flexion-extension: FPL vs. EPL and EPB
Abduction-adduction: APL and APB vs. AddP
Opposition-retropulsion: opponens vs. EPL

3.1 Motion: Kinematics

In the upper limb, most muscles are organized in pairs of **agonists and antagonists** around joints with rotational movements in different directions (degrees of freedom), summarized in Table 3.1. But not all muscles fit in this scheme, and sometimes, there seems to be a physiological imbalance between both muscle groups, when one action is assumed by one single muscle (like shoulder medial rotation by the subscapular muscle) while the antagonistic activity is shared between multiple muscles (the lateral rotators are the infraspinatus, the anterior deltoid, and the teres minor).

Whereas the shoulder glenohumeral joint may be represented by a simple ball-and-socket joint, both the elbow and wrist joints are much more complex and their biomechanical behavior is interdependent (proximal and distal radioulnar joint).

Motion in the upper limb also has to be seen like a linked chain sequence, where the rotational position of the humeral head will influence the prosupination of the forearm.

Any biomechanical consideration serves the local problem under study (for example, the radial head positioning) and thus might be well detailed to understand better and analyze the topic, but for a more global, target-oriented view, simplifications are frequently mandatory.

Motion analysis of the upper limb has been a valuable research tool since more than 20 years, and data collection may be very detailed (Salvia et al., **unpublished**), but the surgically relevant parameters sometimes need to be redefined, as too much data do not allow us a pragmatic view.

3.2 Force Vectors

Textbooks of functional anatomy describe the force vectors related to limb muscles, showing their pull activity line in space. We learned a lot of upper limb motion and especially the muscular balance at the shoulder joint level while studying the "old" anatomic textbooks of von Lanz and Wachsmuth in their reprinted version [1] and classic works on biomechanics [2].

As Aachen was the workplace of the famous German orthopedic surgeon Friedrich Pauwels, we learned how to apply force vectors onto the glenohumeral joint, comparable to Pauwels' extensive biomechanical work considering the hip joint (Fig. 3.1) [3, 4]. In my thesis, I was able to show how daily movements with a medially rotated arm developed forces at the level of the glenohumeral joint responsible for a progressive posterior subluxation of the humeral head, thereby explaining by a biomechanical consideration the known clinical observation of a posterior subluxated humeral head, associated with a medial rotation contracture, in the first years of life in a child suffering from obstetric brachial plexus palsy (see Sect. 2.2; Fig. 3.2).

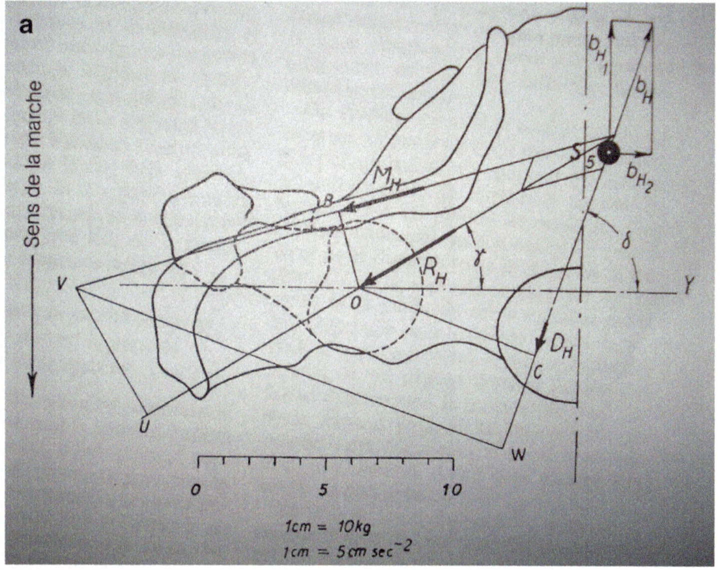

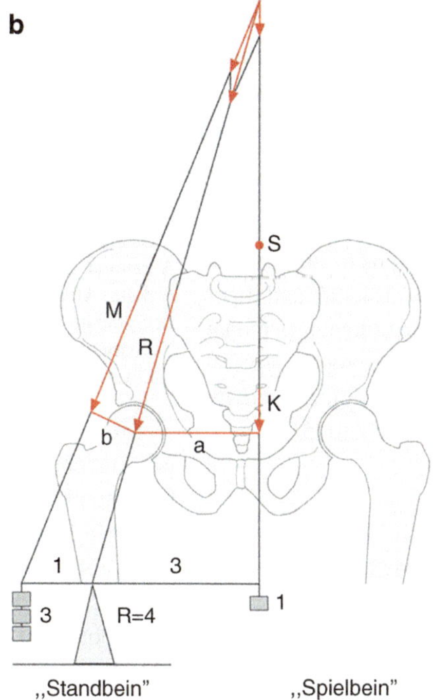

Fig. 3.1 Force vectors around the hip joint, seen: (**a**) in a transverse section (**b**) in an horizontal (frontal) section

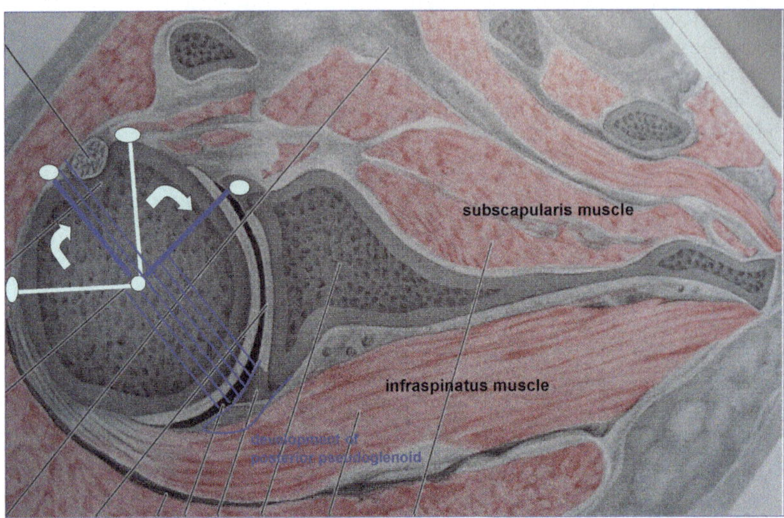

Fig. 3.2 Example of force vectors around the shoulder joint

3.3 Use

We may describe the use of an upper extremity according to its integration into the activities of daily living (body hygiene, dressing, combing, eating or food preparing, basic homework) or considering the complex concept of the "International Classification of Functioning" (ICF) regarding the objectively measurable dimensions of human life. Among the five domains (body structure and function, activity, participation, environment, personal factors), we focus especially on the topics *participation* (social contact, school, family) and *activity* (self-care, play, learning) (Becher T in [5]).

3.4 Basics on Motor and Sensitive Function

Table 3.2 reminds us of the basic peripheral nerve fiber types.

Motor function is assumed by A alpha fibers, and sensation mainly by A beta and delta and C-type fibers. Tuning of motor function by muscle spindles or tendon organs is regulated by A alpha (afferent) and A gamma (efferent) fibers. There is thus no pure motor nerve; even the suprascapular nerve aimed to innervate the supra- and infraspinatus muscle contains sensitive fibers, necessary to regulate the muscle tension via the muscle spindles and tendon organs (Golgi apparatus).

Table 3.2 Peripheral nerve fiber types

Type	Diameter	Conduction velocity (NCV)	Myelinated
A	Large	High	Yes
B	Smaller	Lower	Yes
C	Small	Low	No

Role

A alpha: Efferent to skeletal muscle, afferent from muscle spindles and tendon organs
A beta: Afferent from skin (mechanoreceptors)
A gamma: Efferent to muscle spindles
A delta: Afferent from skin (temperature, "fast" pain)
B: Sympathetic (preganglionic)
C: Sympathetic (postganglionic afferent); polymodal afferent from skin ("slow" pain, thermal, chemical, mechanical)

Motor function needs fine-tuning, achieved by proprioception and probably more sophisticated, but actually poor known mechanisms.

Table 3.3 reminds us of the basic muscle fiber types and their individual properties—any given muscle, according to its particular task, contains a proportional mosaic of those fiber types in order to best achieve the desired function [6].

This means also that when the task of a muscle changes, e.g., after a muscle transfer, the muscle fiber type composition gets arranged in a new way, and this is a process which may last for a year, which will of course influence the receptivity to physiotherapy and development of adequate strength during that adaptation period. There is thus specificity in the reinnervation process from specific motor neurons into the target muscle [7] but also modulation according to the dedicated muscle effector function.

Table 3.3 Muscle fiber types

Type	I (SO)	IIA (FOG)	IIX (FG)
Contraction time	Slow	Fast	Very fast
Resistance to fatigue	High	Intermediate	Low
Activity	Aerobic	Long-term anaerobic	Short-term anaerobic
Motor neuron	Small	Large	Very large
Twitch	Slow (S)	Fast (F)	Fast (F)
Use	tonic	Tonic/phasic	Phasic
Color	Dark	Light	White

Table 3.4 Skin receptors

Type	Reacting to	Located	Name
PC Pacinian corpuscle	Acceleration (vibration)	Subcutis dermis	Vater-Pacini Krause
RA rapidly adapting	Velocity	Papillary dermis	Meissner
SA slowly adapting I	Pressure, traction (vertical)	Basal epidermis	Merkel
SA slowly adapting II	Stretch	Reticular dermis	Ruffini

Interestingly, every motor end plate is also deserved by sympathetic nerve endings, with unknown role, suggesting that a fight or flight scenario's stress might affect pretension-related muscles.

In Table 3.4, we shortly present the most known sensory end organs, classified accordingly to their speed of reaction to various skin stimuli. Up to now, our sensitive peripheral nerve surgery is unable to address specifically those end organs. Also our neuroma surgery remains aspecific, related to A delta and C nerve fibers. Nevertheless, we start to learn more about small fiber (C type) neuropathy (burning pain in lower extremities in diabetic or rheumatoid patients) and neurologists ask us for specific skin biopsy (looking for epidermal nerve fiber density) to rule out this diagnosis.

3.5 Cortical Plasticity

The concept behind is probably better named "neural plasticity." It is a huge and ever-growing field of knowledge showing some ways of how the peripheral and central nervous systems are able to adapt and learn, starting from the basic regenerative processes of sprouting and pruning up to the complex changes in cortical representation in motor and sensation areas following nerve trauma, central nervous diseases, etc.

Cortical plasticity is a wonderful concept, describing our neocortex capacity of learning and adaptation. One of the first images coming into our mind is Penfield's homunculus and its imaged idea of the cortical representation zones of various body parts in the sensory and motor cortex and their relative importance, in terms of topography, size, and density.

On such a scheme, we may understand Göran Lundborg's elaborated explanations about the processes happening when sensory or motor nerves get injured and transected—summarized by the loss of a representation area and its progressive and suboptimal restitution once nerve regeneration moves on [8]. We are aware that adjacent body parts (like fingers) have adjacent topographic representation areas, and we explain changes in sensation and motor control referring to those mechanisms.

Those adaptive processes are based on neurophysiological learning and memory processes, which another genius neuroscientist discovered and described so well: Nobel Prize winner Eric Kandel did most of his studies of neurophysiology of memory in a marine mollusk with only 20,000 neurons, some of them being as giant as 1 mm of diameter (*Aplysia californica*). In his textbook on neuroscience, first published in 1981 [9], and in an extended novel telling his life and scientific carrier [10], he describes the molecular and genetic changes associated with short- and long-term memory and more specifically the different learning forms of habituation, sensitization, and conditioning. His books are a fascinating life story recommended for every interested scientist or physician.

Years before, the American neurologist Frank Wilson wrote an amazing comparative study about the hand in human evolution: "How Its Use Shapes the Brain, Language, and Human Culture" [11].

Cortical plasticity thus may be understood on a molecular, cellular, and structural tissue level involving physiological and pathological processes related to brain activity. Motor and sensory control are not the only ones, but those interesting the hand (and peripheral nerve) surgeon.

In our practice, we get more and more interested in nerve transfers and treatment ways to alleviate neuropathic pain.

Nerve transfers bring healthy donor fascicles onto sensory or motor targets who were denervated by previous peripheral or central nerve injury, illness, or tumors. This technique had been described in 1913, over hundred years ago, by a German colleague Alfred Stoffel, but started to get popularized only in the 1990s. One of the first and most used motor nerve transfers is today named "Oberlin's transfer," and it reneurotizes the biceps and/or brachialis muscle by redundant motor fascicles coming from the unaffected median and/or ulnar nerve [12]. After the nerve transfer, the new relevant motor area in the cortex related to the motor donor is rewired for the new function [13, 14].

Meanwhile, their use is progressively extended from peripheral nerve injury cases to medullary trauma (para- and tetraplegia) or infection (acute transverse myelitis) and even central nerve conditions like stroke (spastic hemiplegia) and spasticity (contralateral root C7 transfer, which decreases the hypertonia by the ipsilateral C7 deconnection and adds controllable new motor input by the other healthy C7; [15]). It is interesting to imagine that in cases of upper limb spasticity, the surgical reduction of motor input known as hyponeurotization or hyperselective neurotomy initially described by a famous hand surgeon, Giorgio Brunelli, nowadays is refined [16] and may be extended to the addition of gearing new motor nerve transfers from adjacent but healthy, non-spastic donors, regulated by voluntarily controllable cortical areas.

Also does the more extended use of contralateral nerve transfers (mostly the rather dispensable C7 root) teaches us that the cortical reorganization may overcome laterality concerns, as for proximal upper limb tasks, a frequent parallel and simultaneous bilateral activation preexists already (lifting of heavy objects involves bilateral shoulder and biceps activation, and even addition of muscles like the intercostals or diaphragm not directly involved in the specific movement, thereby illustrating also how, e.g., intercostal nerves innervating the accessory breathing muscles may successfully be used to reanimate elbow flexion power using a nerve transfer onto the motor nerves to brachialis and biceps muscle) [17].

Those surgical techniques, based on a growing knowledge of our cortical reorganization, are already extended to totally different fields of reconstructive surgery, like urology (neurogenic bladder, [18]) or ophthalmology (corneal resensibilization, [19]). We successfully started to treat children with upper limb arthrogryposis using nerve transfers to reanimate elbow flexion and shoulder lateral rotation [20]. It is up to us experienced with this technique to address other specialties and seek for new indications, which start to be "at hands."

There are also rehabilitation concepts addressing cortical plasticity in hand surgery, especially considering the sensitive and nociceptive pathways. The well-established mirror therapy and graded motor imagery [21, 22] are very useful in the treatment of neuropathic pain and CRPS. Motor nerve transfers in children do not need a specific target therapy, as children are able with some months' delay to address the newly gained function. This applies for classic motor nerve transfers in the upper limb, like the Oberlin or Somsak transfer. Also intercostal nerve transfers aimed to reanimate elbow flexion and extension work well.

The rehabilitation is more delicate in adults, who need more time and an exercise program where the motor function of the donor area (like finger flexion in an Oberlin transfer, breathing or coughing when the intercostal nerves have been transferred) must be addressed to start biceps activity, which will need a parallel trigger training for months, before the targeted movement may be addressed separately (active elbow flexion becoming addressable even with extended fingers).

Although a lot of basic knowledge is available [23], those techniques need to be better integrated in training programs after functional reconstructive surgery.

This does not mean that cortical plasticity has only an interest to explain why some of our surgical strategies work—the concept itself is so beautiful and challenging that all higher cerebral functions including specific mental states like coma or near-death experiences are to be investigated and start to deliver fantastic insights [24].

Moving further into those concepts will provide more indications and surgical solutions to conditions otherwise untreated or neglected. It is therefore worth not only to apply and refine a microsurgical technique, but also to participate in the global thinking of the paradigm behind the scene and to work on networks between so distant disciplines as those the here-cited authors represent in science.

References

1. von Lanz T, Wachsmuth W. Praktische Anatomie (Arm). Berlin: Springer; 1959.
2. Kummer B. Biomechanik—form und Funktion des Bewegungsapparates. Köln: Deutscher Ärzteverlag; 2005.
3. Pauwels F. Gesammelte Abhandlungen zur funktionellen Anatomie des Bewegungsapparates. Berlin: Springer; 1965.
4. Pauwels F. Atlas zur Biomechanik der gesunden und kranken Hüfte. Berlin: Springer; 1973.
5. Bahm J, editor. Movement disorders in children's upper limb. Springer; 2020.
6. Qaisar R, Bhaskaran S, Van Remmen H. Muscle fiber type diversification during exercise and regeneration. Free Radic Biol Med. 2016;98:56–67.
7. Madison RD, Robinson GA, Chadaram SR. The specificity of motor neuron regeneration (preferential reinnervation). Acta Physiol (Oxf). 2007;189(2):201–6.
8. Lundborg G. Nerve injury and repair: regeneration, reconstruction and cortical remodeling. Churchill Livingstone; 2005.
9. Kandel ER, Schwartz JH, Jessell TM. Principles of neural science. Elsevier; 1981.
10. Kandel ER. In search of memory—the emergence of a new science of mind. 2007.
11. Wilson FR. The hand: how its use shapes the brain, language, and human culture. 1999.
12. Teboul F, Kakkar R, Ameur N, Beaulieu JY, Oberlin C. Transfer of fascicles from the ulnar nerve to the nerve to the biceps in the treatment of upper brachial plexus palsy. J Bone Joint Surg Am. 2004;86(7):1485–90.
13. Anastakis DJ, Malessy MJ, Chen R, Davis KD, Mikulis D. Cortical plasticity following nerve transfer in the upper extremity. Hand Clin. 2008;24(4):425–44, vi–vii.
14. Socolovsky M, Malessy M, Lopez D, Guedes F, Flores L. Current concepts in plasticity and nerve transfers: relationship between surgical techniques and outcomes. Neurosurg Focus. 2017;42(3):E13.
15. Zheng MX, Hua XY, Feng JT, Li T, Lu YC, Shen YD, Cao XH, Zhao NQ, Lyu JY, Xu JG, Gu YD, Xu WD. Trial of contralateral seventh cervical nerve transfer for spastic arm paralysis. N Engl J Med. 2018;378(1):22–34.
16. Leclercq C, Perruisseau-Carrier A, Gras M, Panciera P, Fulchignoni C, Fulchignoni M. Hyperselective neurectomy for the treatment of upper limb spasticity in adults and children: a prospective study. J Hand Surg Eur. 2021;46(7):708–16.
17. Malessy MJ, Bakker D, Dekker AJ, Van Duk JG, Thomeer RT. Functional magnetic resonance imaging and control over the biceps muscle after intercostal-musculocutaneous nerve transfer. J Neurosurg. 2003;98(2):261–8.
18. Agarwal P, Parihar V, Kukrele RR, Kumar A, Sharma D. Anatomical feasibility of anastomosing intercostal nerves (D10&D11) and subcostal nerve (D12) to S2 ventral root and lumbar plexus for management of bladder function after spinal cord injury. J Clin Orthop Trauma. 2020;11(5):900–4.
19. Catapano J, Fung SSM, Halliday W, Jobst C, Cheyne D, Ho ES, Zuker RM, Borschel GH, Ali A. Treatment of neurotrophic keratopathy with minimally invasive corneal neurotisation: long-term clinical outcomes and evidence of corneal reinnervation. Br J Ophthalmol. 2019;103(12):1724–31.
20. Bahm J. Arguments for a neuroorthopaedic strategy in upper limb arthrogryposis. J Brachial Plex Peripher Nerve Inj. 2013;8(1):9.
21. Moseley GL. Graded motor imagery is effective for long-standing complex regional pain syndrome: a randomised controlled trial. Pain. 2004;108(1–2):192–8.
22. Strauss S, Barby S, Härtner J, Pfannmöller JP, Neumann N, Moseley GL, Lotze M. Graded motor imagery modifies movement pain, cortical excitability and sensorimotor function in complex regional pain syndrome. Brain Commun. 2021;3(4):fcab216.
23. Moucha R, Kilgard MP. Cortical plasticity and rehabilitation. Prog Brain Res. 2006;157:111–22.
24. Lane R. Steven Laureys: a clinical focus on consciousness research. Lancet. 2020;396(10264):1719.

4 Invited Contributions: Fields Outside of My Work

J. Bahm, *Surgical Rationales in Functional Reconstructive Surgery of the Upper Extremity*,
https://doi.org/10.1007/978-3-031-32005-7_4

4.1 Surgical Techniques in Upper Limb Amputation

By Gregor Laengle, Clemens Gstoettner, Oskar C. Aszmann from Vienna, Austria.

The surgical reconstruction of hand function is a complex procedure, which involves certain principles and considerations to achieve an optimal outcome. The limited possibilities of biologic hand reconstruction can be expanded by replacing the lost hand with high-tech myoelectric hands, which are driven by biologic signals of the remaining neuromuscular units. These modern technologies experience a continuous development and are topic of numerous research endeavors. As of today, the standard prosthetic fitting of a bionic prothesis consists of an individually adapted socket that is either directly attached to the stump via negative pressure or with the aid of additional harnesses fixed to the shoulder and trunk. These sockets include electrodes at specific positions to sense electromyographic activity on the skin surface of the residual limb. At least two stable signals are needed for reliable and functional prosthetic control. Depending on the requirements and expectations of the patient, different hand devices are available, starting from simple and robust grippers to fine and aesthetic hands with multiple degrees of freedom and single-finger movements. Next to the rather simple body-powered prosthesis, which utilizes cable-driven mechanical movements from the shoulder joint, there is a notable progress of myoelectric technologies towards a natural prosthetic experience. Implantable technologies directly interface muscles or nerves from the stump to establish a more stable and accurate signal exchange.

The concept of bionic hand reconstruction follows an interdisciplinary approach of doctors, physiotherapists, orthopedic technicians, psychologists, and others to warrant a desirable outcome. This can be best achieved with specialized centers which offer a comprehensive patient treatment plan, from surgery to prosthetic fitting and rehabilitation.

The ability to artificially mimic hand function through sophisticated devices is a remarkable achievement of the last decades. The positive integration of a bionic hand into the body image has successfully proven to increase amputee's quality of life. Nevertheless, the technical approximation is always inferior to biologic alternatives. Therefore, the surgeon should prioritize biologic over bionic reconstruction whenever possible.

4.1.1 Patient Identification

There is only limited information on the prevalence of patients suffering limb amputation. The overall number of patients living with limb loss in the United States was estimated to be 2.2 million people for the year 2020, one-third accounting for the upper limb [1]. Whereas lower limb amputations are more common due to the underlying cardiovascular diseases, hand amputation usually occurs through a traumatic injury or oncologic surgery. Restoring the upper extremity is a challenging task, which requires not only expertise from the treating personnel, but also full commitment and adherence from the patient. Since patients with upper limb amputation tend to be younger and healthier than those with lower limb loss, patient compliance and motivation are generally favorable. Nevertheless, care should be taken to ensure that the patient's mental and physical as well as psychosocial status is appropriate to the following rehabilitation process, which may be demanding [2].

The need for surgical intervention in prosthetic upper limb replacement focuses on improving the socket attachment and enhancing the signal quantity or quality for prosthetic control. The upcoming technique of bone anchorage of the prosthesis (osseointegration) is of great value to certain conditions, where classical attachment limits the use of the device (e.g., transhumeral level). Moreover, correction of the soft tissues of the stump is a common indication for surgical referral. If the residual limb does not offer reliable signals, selective nerve transfers (targeted muscle reinnervation) can be used to create several new electromyographic signals. In certain cases, the transplantation of a free muscle to the amputation site is needed.

The reconstruction of a functional hand with a bionic prosthesis is not only applicable to patients with hand loss but may also help individuals with irreversible damage of the soft tissues or brachial plexus lesions. Those patients suffering from the miserable state of a useless hand without any relevant motor or sensory function may be evaluated for an elective amputation and consecutive fitting of a myoelectric prosthesis. This therapeutic option can help reduce pain and regain capability for everyday tasks, which has been demonstrated in various studies [3, 4].

4.1.2 Amputation Level

4.1.2.1 Minor Amputations

The reconstructive options depend on the level of amputation. Minor amputations comprise the loss of a single or multiple digits as well as partial hand amputations. These situations can often be handled by attempts of finger replantation or secondary procedures like pollicization of the index finger or free transfer of the second toe. Amputation of the thumb causes the greatest impairment of hand function and thus is most relevant in reconstruction. Prosthetic reconstruction of single fingers focuses on the aesthetic aspects. However, silicone prostheses show functionality to a certain degree. In cases of short phalangeal or metacarpal bone stumps, the use of osseointegrative implants can facilitate prosthetic attachment.

4.1.2.2 Major Amputations

Transcarpal

The transcarpal level is characterized by complete preservation of the radius and ulna. Hence, patients are still able to perform pro- and supination. However, in terms of prosthetic fitting, the length of the forearm is challenging for acquiring a symmetric body image since additional arm length will be added due to the prosthetic components, leading to an undesired appearance. Thus, the option of transradial shortening of the stump is a sensible adaption facilitating prosthetic fitting without loss of functionality.

Transradial

This level of amputation is very frequent and highly impacts a patient's overall ability and occupational situation. An optimal stump length of 16 cm from the lateral epicondyle assures enough space for prosthetic components and rotational movements of the forearm. A minimum of 5 cm is necessary for a functional elbow flexion [5]. The remaining muscles and soft tissues usually allow good osseous coverage and tight stump adaption for pain prevention; however, neuroma pain is a common problem at this amputation level [6, 7]. Ideally, prophylactic surgical procedures during primary amputation, like intramuscular nerve transposition, prevent the development of neuroma pain.

At forearm level, there are usually enough muscular signals to control a prosthesis, and thus only rarely are surgical interventions like selective nerve transfers or free functional muscle transfers indicated. Prosthetic fitting can generally start early after operation (3–4 weeks) due to quick recovery, compared to the lower extremity.

4.1.2.3 Elbow Exarticulation

Elbow exarticulation is a rare level of amputation, which shows particular features for prosthetic reconstruction. The preservation of the humeral condyles can improve prosthetic suspension due to the cone-shaped bone and hence obviate the application of bothering straps. Moreover, rotational movements of the glenohumeral joint can be directly translated. The preserved length of the humerus is unfavorable for customizing a prosthetic elbow joint, since the overlength of the upper arm has an unnatural appearance and may also cause difficulties in usage. Segmental shortening of the humerus unites the advantages of prosthetic attachment with a shorter limb length appropriate for prosthetic fitting [8].

Transhumeral

Next to forearm amputations, transhumeral is the most frequent major amputation level of the upper extremity [9]. One major issue with proximal amputation levels is the number of degrees of freedom that need to be substituted (elbow, wrist, fingers). At transhumeral level, patients require several muscular signals for prosthetic control, while usually only two are available (biceps and triceps). To increase signal quantity, targeted muscle reinnervation (TMR) offers a viable option. Specifically, the major nerves of the hand (median, ulnar, radial) are rerouted to the remaining stump muscles (biceps, triceps) to generate additional EMG signals that can be distinctively controlled.

For prosthetic attachment, additional harnesses are essential to secure the device to the body; this, however, compromises range of motion in the shoulder. Such solutions have shown to be uncomfortable for patients and result in frequent abandonment [10]. To improve prosthetic attachment, a humerus flexion osteotomy can enhance prosthetic stability and rotational movements [5]. Alternatively, osseous fixation of the prosthesis brings great advantages to the user, since it preserves full range of motion in the shoulder in contrast to standard models with straps, while simplifying donning and doffing of the prosthesis. This makes transhumeral patients ideal candidates for osseointegrative prosthesis [11]. In very short transhumeral amputees, the stump length does not provide enough surface to allow a standard prosthetic attachment. This situation requires a socket fitting, which encloses the entire shoulder girdle, rendering the patient a glenohumeral amputee from a functional perspective, with total loss of shoulder joint motion. With osseointegration, however, such patients can retain full abduction and rotation in the shoulder, given that the respective muscles are unharmed.

4.1.2.4 Glenohumeral/Forequarter

The absence of the entire arm is equal to a shoulder exarticulation. This level is associated with the greatest impairment of function and patient dissatisfaction due to the problems with prosthesis attachment. Available weight-bearing anatomical structures are solely the acromion, clavicula, and trunk. Due to lack of appropriate bony fixation, these patients are generally not candidates for osseointegration. There is no better alternative for socket suspension than using belts that cover not only the affected shoulder but also the upper part and contralateral side of the trunk, which is highly inconvenient. EMG signals can be detected from the trunk musculature (pectoral major and minor, latissimus dorsi, supra- and infraspinatus). These patients may also profit from TMR surgery since the neuroanatomy of the pectoral muscles offers separate targets for reinnervation [12].

In very rare cases, referred to as forequarter amputation, the clavicula and scapula need to be removed as well, often due to malignant tumors. Surgical ablation often leads to soft tissue defects, which can be closed with either fillet flaps of the amputated limb or free flaps (e.g., tensor fasciae latae flap) (Fig. 4.1) [13].

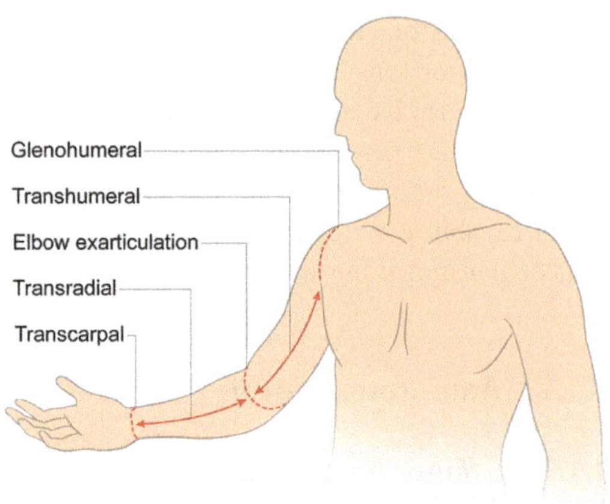

Fig. 4.1 Illustration of major amputation levels. © Aron Cserveny 2022

4.1.3 Surgical Techniques and Indications

4.1.3.1 Nerves

Targeted Muscle Reinnervation

Targeted muscle reinnervation is one of the major advances in prosthetic care in recent decades, bringing various improvements to the prosthetic user. This technique was primarily postulated in 1980 and was first introduced in a patient with bilateral shoulder disarticulation and conceptualized for proximal amputations [14, 15]. In glenohumeral and transhumeral patients, there is an increase of degrees of freedom that need to be substituted, while only few myosignals are available. In these cases, the remaining stump muscles can be reinnervated to create new myosignals. The idea of exploiting otherwise inactive nerves of the stump in above-elbow amputation revolutionized the possibilities for prosthetic control. Nowadays, it is a well-proven procedure and surgical techniques have been described for glenohumeral, transhumeral, transradial, and lower extremity amputations [7, 16–18].

TMR is indicated in patients with dissatisfying prosthetic handling, due to weak or insufficient EMG signals and can be performed as a secondary reconstruction or at primary amputation. Since below-elbow amputations usually do not lack EMG signals, this procedure is mainly employed for above-elbow amputations.

Specific anatomical knowledge of the brachial plexus at different topographic levels is vital to maintain a clear overview of the neuronal structures. A thorough preoperative workup is essential to classify the neuromuscular status of the patient. Clinical examination guides the identification of damaged peripheral nerves. High-resolution ultrasound and magnetic resonance imaging (MRI) help confirm the extent of the neuronal damage or nerve root avulsion. Electrophysiologic studies are beneficial in situations of subclinical motor nerve function; however, the use of this method is very limited in proximal amputations.

Specific nerve transfer matrices have been proposed for standard surgical treatment in TMR (see Tables 4.1 and 4.2). Hereby, it is possible to acquire up to six separate signals, resulting in three degrees of freedom for standard myoelectric control.

In transhumeral cases, the effector muscles are the flexors and extensors of the upper arm (biceps, brachialis, triceps). Depending on the extent of nerval lesions, it might be necessary to do a supraclavicular plexus exploration and reconstruction to repair severed fascicles which contain relevant donor nerves. The target muscles are accessed with a longitudinal incision from the anterior axillary fold, extending along the medial bicipital groove distally. The musculocutaneous, median, and ulnar nerves are exposed, and the muscular branches of the biceps and brachial muscles are prepared. During this procedure, the short head of the biceps muscle should be separated from the long head and ideally an adipose tissue flap is inserted to reduce electromyographic cross talk between the two muscles. After careful preparation of the muscular branches, intraoperative nerve stimulation confirms the functionality of the muscles. Since nerve stumps usually end up in a terminal neuroma, the donor nerves (median and ulnar nerve) are shortened to a point of healthy fascicular architecture before coaptation. The muscular branches are then dissected at a distal level a few centimeters from the muscular entry point to ensure a short regeneration time. Afterwards, the radial nerve is addressed through a dorsal incision and blunt dissection of the triceps lateral and long heads. The lateral head is mobilized to gain access to the motor branch, and nerve coaptation with the deep radial nerve is performed as mentioned before.

Table 4.1 Standard nerve transfer matrix in transhumeral patients

Targeted muscle	Nerves	Prosthetic function	Innervation
Biceps long head	Musculocutaneous	Elbow flexion	Native
Biceps short head	Ulnar	Hand close	Transferred
Brachialis	Median	Pronation	Transferred
Triceps long and medial head	Radial	Elbow extension	Native
Triceps lateral head	Deep radial branch	Hand open	Transferred
Brachioradialis	Split deep radial branch	Supination	Transferred

Table 4.2 Standard nerve transfer matrix in glenohumeral patients

Targeted muscle	Nerves	Prosthetic function	Innervation
Pars clavicularis pectoralis	Musculocutaneous	Elbow flexion	Transferred
Pectoralis minor	Ulnar	Hand close	Transferred
Pars sternocostalis pectoralis	Median	Hand close/wrist rotation	Transferred
Pars abdominalis pectoralis	Median	Wrist rotation	Transferred
Latissimus dorsi	Radial	Elbow extension	Transferred
Infraspinatus	Deep radial branch	Hand open	Transferred

In cases of a long humeral stump, the brachioradialis muscle may provide an additional target for a split deep radial nerve transfer.

In glenohumeral amputations, the pectoralis major and minor muscles are the main muscular targets for the median and ulnar nerves. The latissimus dorsi and supraspinatus muscles serve as effector muscles for the radial nerve. An infraclavicular curvilinear incision is made, and the pectoralis major muscle is undermined from laterally. To get full access to the pectoralis nerves and provide a separate signal, the pectoralis minor muscle is detached from the coracoid as well as the thorax and transposed laterally to separate it from the pectoralis major muscle. The motor branches can then be easily recognized and secured with vessel loops to help maintain overview. The major nerves of the brachial plexus are identified according to the topographic location of the axillary artery. Nerve coaptation is performed according to a predefined pattern, which may be adapted depending on the specific anatomy of each patient. For better electromyographic signal quality, the subcutaneous tissue overlying the target muscles should be thinned while preserving vascular supply. After surgery, it takes about 3 months until the first signals can be expected, and some more months of training to obtain stable and reliable signal quality (Figs. 4.2, 4.3, and 4.4).

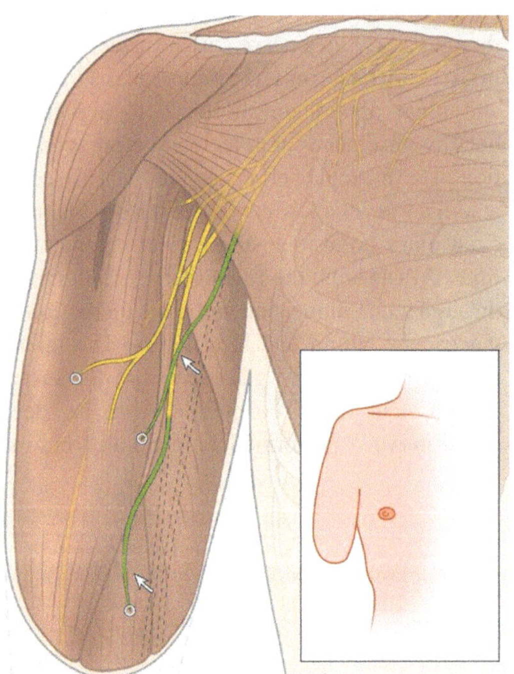

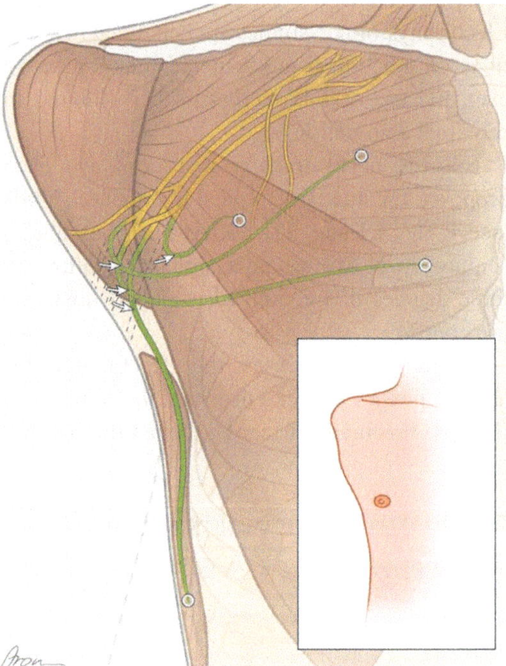

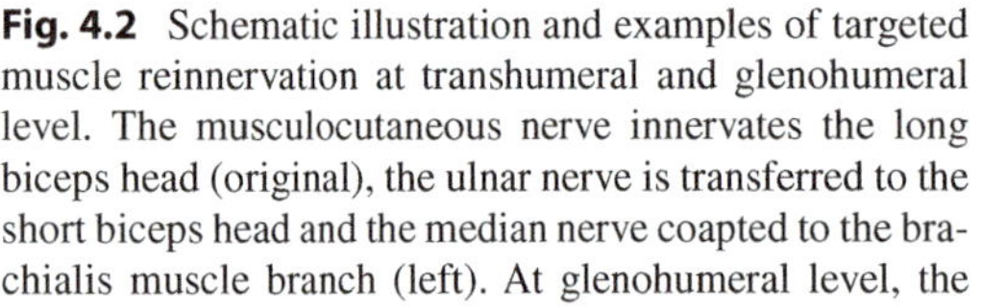

Fig. 4.2 Schematic illustration and examples of targeted muscle reinnervation at transhumeral and glenohumeral level. The musculocutaneous nerve innervates the long biceps head (original), the ulnar nerve is transferred to the short biceps head and the median nerve coapted to the brachialis muscle branch (left). At glenohumeral level, the musculocutaneous nerve is transferred to the clavicular portion, the median nerve to the sternocostal portion of the pectoralis major muscle. The ulnar nerve is rerouted to the pectoralis minor muscle and the radial nerve to the thoracodorsal nerve (right). © Aron Cserveny 2022

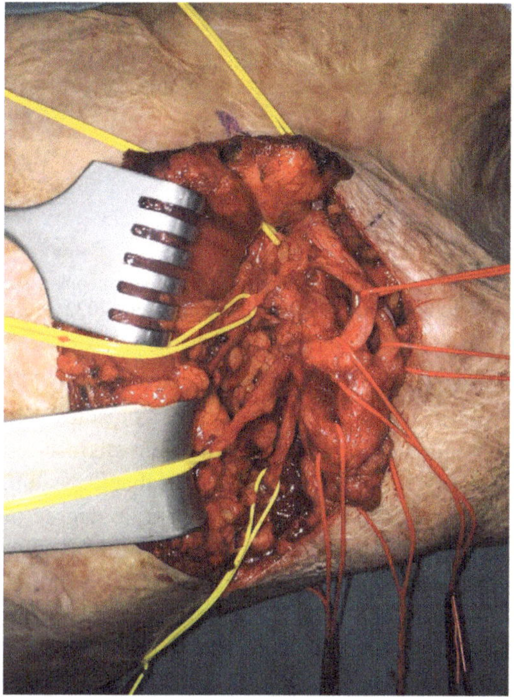

Fig. 4.3 Intraoperative situs of a prepared left brachial plexus of a glenohumeral amputated patient after severe electrocution injury. Donor nerves (musculocutaneous, radial, median, ulnar,) are labelled in red, target nerves to pectoralis major muscle (clavicular, sternocostal, abdominal), latissimus dorsi muscle and serratus anterior muscle are marked in yellow

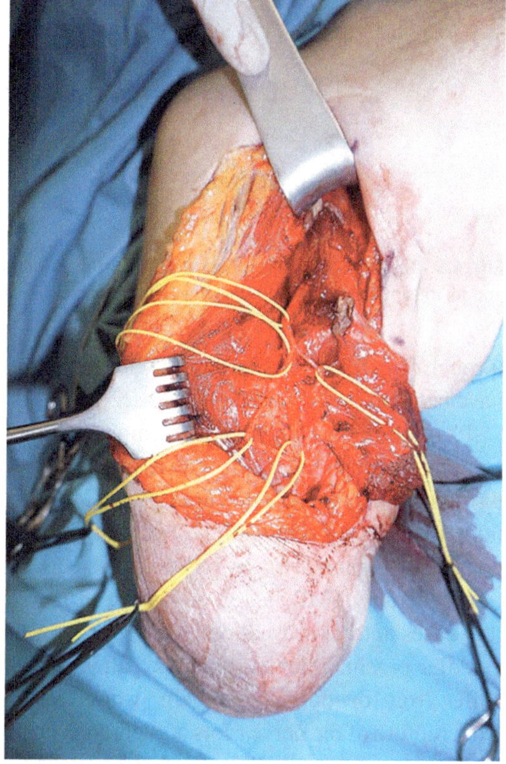

Fig. 4.4 Targeted muscle reinnervation in a transhumeral patient. The vessel loops mark the musculocutaneous nerve and its muscular branches to the short and long bicep head as well as brachialis muscle

4.1.3.2 Neuroma and Pain

After limb amputation, neuroma formation is a common problem and poses a great burden to the patient. Associated neuropathic pain—a form of residual limb pain—often leads to prosthetic abandonment. General prophylaxis comprises covering the nerve stumps with muscular or soft tissue during amputation to prevent undue scarring and chronic stimulation of the terminal afferent fascicles. It may be helpful to perform a shortening and intramuscular transposition of the nerve with an epimysial suture to secure its position. If a neuroma has already formed, neuroma resection and deep, soft burying must be performed. The indication for a surgical correction is always an individual decision, depending on the intensity of pain and limitations in everyday life as well as the findings of clinical examination and imaging.

Numerous techniques for neuroma treatment and prevention have been proposed, and yet no uniformly accepted gold standard exists according to literature [19]. For instance, intraosseous burying, end-to-side neurorrhaphy, centrocentral neurorrhaphy, or neuronal capping with biologic or synthetic material has been described. Especially, in the upper extremity, TMR offers a valuable procedure, which has shown to be effective in treating as well as preventing neuroma and phantom limb pain, while also addressing functional aspects of prosthetic control. Given these benefits, TMR may advance to a standard procedure in primary amputation above the elbow. However, it is also a time-consuming procedure, particularly for surgeons who do not regularly perform peripheral nerve surgery. The main mechanism of TMR in neuroma prevention is the regeneration of the nerve stump into a new target

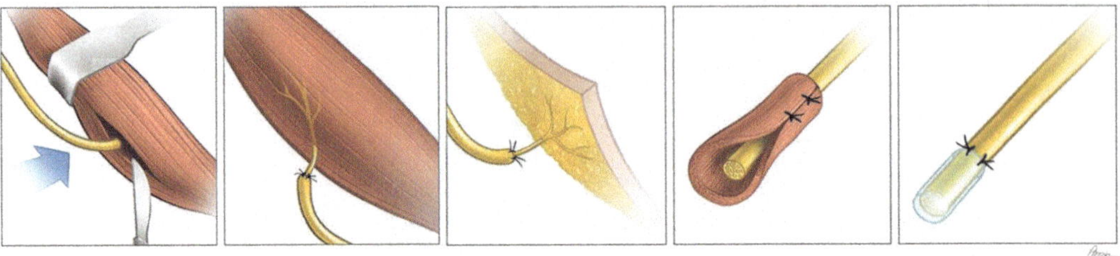

Fig. 4.5 Targeted muscle reinnervation in a transhumeral patient. The vessel loops mark the musculocutaneous nerve and its muscular branches to the short and long bicep head as well as brachialis muscle. © Aron Cserveny 2022

through directed axonal sprouting. Similar effects have been proposed for RPNIs [20–22].

Phantom limb pain is another, more complex form of neuropathic pain, which is thought to have both peripheral and central effects contributing to projected aching of the lost limb. Next to conservative measures, as physio- and occupational therapy, TMR is a good option in refractory cases and for prevention [23].

After brachial plexus avulsion, patients often suffer from deafferentation pain, which results from rupture of the nerve roots at their emergence from the spinal cord. It is characterized by an undulating course with pain attacks, often prompting immediate termination of the current task by the patient. This situation is generally not amenable to surgical intervention, and drug therapy with anticonvulsives, antidepressants, and opioids constitutes the main treatment. However, the pain attacks usually get less frequent and severe over the years (Fig. 4.5).

4.1.4 Emerging Surgical Concepts

Recently, a technique has been described to ameliorate electromyographic signals in amputees by putting small avascular muscle grafts around the end of peripheral nerves. This so-called regenerative peripheral nerve interface (RPNI) forms small neuromuscular units, which generate electric signals that can be selectively controlled and deciphered through myoelectric electrodes. This promising method for prosthetic control is yet to be implemented into clinical routine [24, 25].

Another strategy that focuses more on proprioceptive feedback is called the agonist-antagonist myoneural interface [26, 27]. Herein, pairs of antagonistic muscles are dynamically connected via their disinserted tendons to create a natural feedback system. Each muscle pair stands for a single degree of freedom substituted by the prosthesis. The contraction of one muscle leads to a subsequent elongation in its counterpart, where proprioceptive organs are then stimulated. With this construct, the patient is able to progressively control the muscle activity, which is then translated into prosthetic movement. This method was mainly investigated in lower limb amputation models, but further studies will assess its suitability in the upper extremity.

4.1.5 Soft Tissues

The soft tissues of the stump need to be properly aligned with the right tension. Whereas in the lower extremity this is even more important, in the upper extremity, the prosthetic socket does not receive heavy loads. Several weeks after surgery, the swelling of the stump subsides and excess skin might be problematic for socket adaption.

Residual muscles and tendons should serve as a strong deep layer and cover for bone prominences that would otherwise cause pressure marks. Myoplasty (connecting two different muscle ends) and myodesis (attaching muscle to bone) are approbate means for creating a solid, resilient stump. Skin closure should preferably be done with a volar flap since there is usually less hair growth, which might interfere with wound healing. At the level of the axilla, sweat glands should be excised to avoid development of folliculitis under the socket.

Other indications for surgical soft tissue correction are contractures above joints (elbow, shoulder), e.g., due to chronic muscular disbalances. In cases where physiotherapy is not effective, the contract muscles or tendons can be elongated or even disinserted as a therapeutic and prophylactic treatment. Scar tissue of the skin can be excised, and basic skin plasties can be performed (Z-plasty, etc.).

4.1.6 Bones

Usually, the osseous structures do not need special treatment. During primary amputation, it is important to round off sharp bone edges. Several techniques for bone treatment exist that generally focus on the suspension of the prosthesis.

4.1.7 Segmental Shortening

In the special case of elbow exarticulation, the humeral epicondyles offer an ideal support for prosthetic attachment due to the cone-shaped form. An additional advantage is the improved ability to perform rotational movements. Hence, a segmental shortening of the humerus can be performed to create enough space for an elbow joint at a symmetric level as on the contralateral side. During this procedure, it is also possible to do a combination with a rotational osteotomy, if internal contracture in the shoulder joint is present. Derotation of the humerus alone may also be performed, especially in patients with brachial plexus lesions.

4.1.8 Angulation Osteotomy

To improve prosthetic attachment in long transhumeral amputations, angulation osteotomy of the humerus can provide an alternative in cases of insufficient or cumbersome prosthetic suspension. With this technique, the humerus is cut distally and the distal part is reorientated ventrally in an angle of 70–110°. Fixation can be done with Kirschner wires, compression screws, or an osteosynthesis plate. This artificial angle is prone to straighten out in about one-third of cases. This technique might not only reduce additional harnesses, but also enhance rotational control. Still, the indication is mainly reserved for children and rarely performed in adults.

4.1.9 Osseointegration

Osseointegration describes the full integration of an intramedullary metal implant, which can be used for prosthetic anchorage. Particularly in very proximal transhumeral amputees, osseointegration displays a valuable alternative for a convenient prosthetic attachment, since conventional suspension often leads to discomfort and stump problems, including signal instability, skin irritation, and pain [11]. Currently, there are two press-fit systems (integral leg prosthesis, ILP; osseointegrated prosthetic limb, OPL) and one thread-based system (osseointegration prosthesis for rehabilitation of amputees, OPRA) available. The author's experience is related to the latter, which is suitable for both upper and lower extremites. The implantation generally consists of a two-stage procedure with an interval of 3–6 months. In the upper extremity, this can also be done in a single surgery. During the implantation, an intramedullary fixture is inserted and connected to a percutaneous abutment for socket attachment. The cutaneous port is created by punching a circular hole through a preformed skin flap, which is freed from subcutaneous tissue, so the skin directly adheres to the bone. This facilitates dry and clean wound conditions without much secretion.

The osseous fixation of the prosthesis has several advantages compared to cumbersome belts that envelop the contralateral side of the trunk. The mobility and range of motion with osseointegration devices are superior to conventional alternatives. Moreover, osseointegrated devices have been reported to enable users to better sense vibrations from the prosthesis as a form of feedback mechanism, termed osseoperception [28].

The percutaneous port of the bone-anchored implant can be further utilized for inserting implantable electrodes to interface the stump musculature or nerves. Hence, the trend to implantable technologies, which yield more accurate and stable man-machine interaction, can be pursued [29]. Nevertheless, the increased risk for minor infections must be considered and the patient needs to be instructed to hygienic measures (Figs. 4.6, 4.7, 4.8, 4.9, and 4.10).

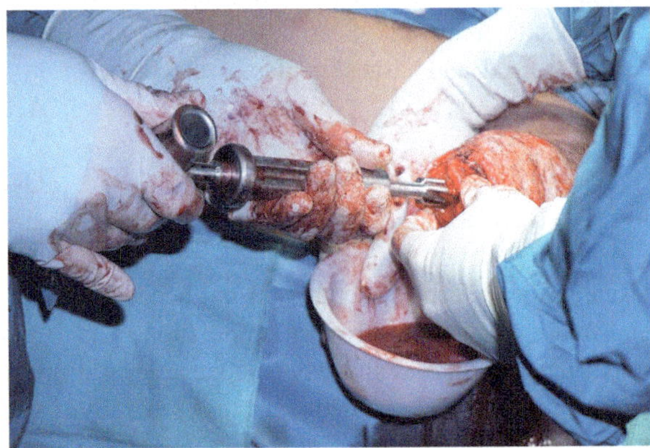

Fig. 4.6 Intraoperative image during transhumeral osseointegration. The osseous canal is manually drilled to prepare insertion of the intramedullary fixture

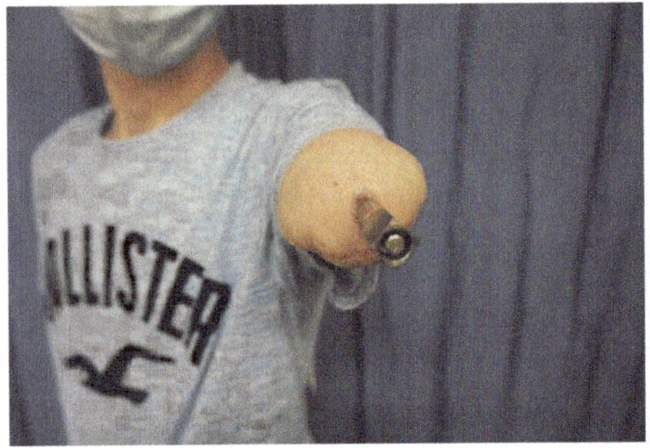

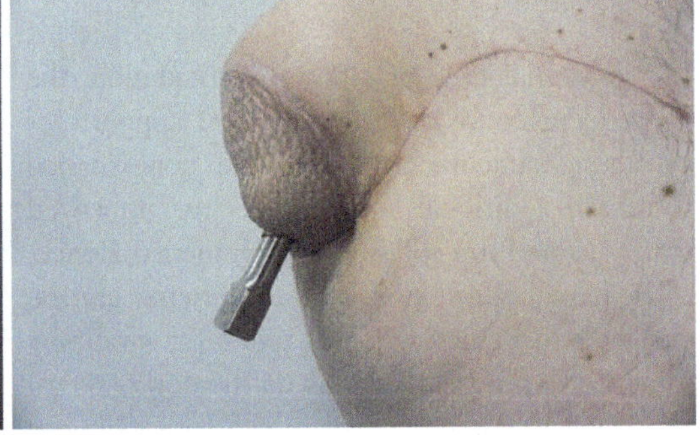

Figs. 4.7 and 4.8 Clinical appearance after transhumeral osseointegration in a long and short humeral stump with special regard to the cutaneous port. The skin is thinned and adhered to the bony end to create a solid stump and prevent secretion. The short stump has been primarily covered with split-thickness skin graft

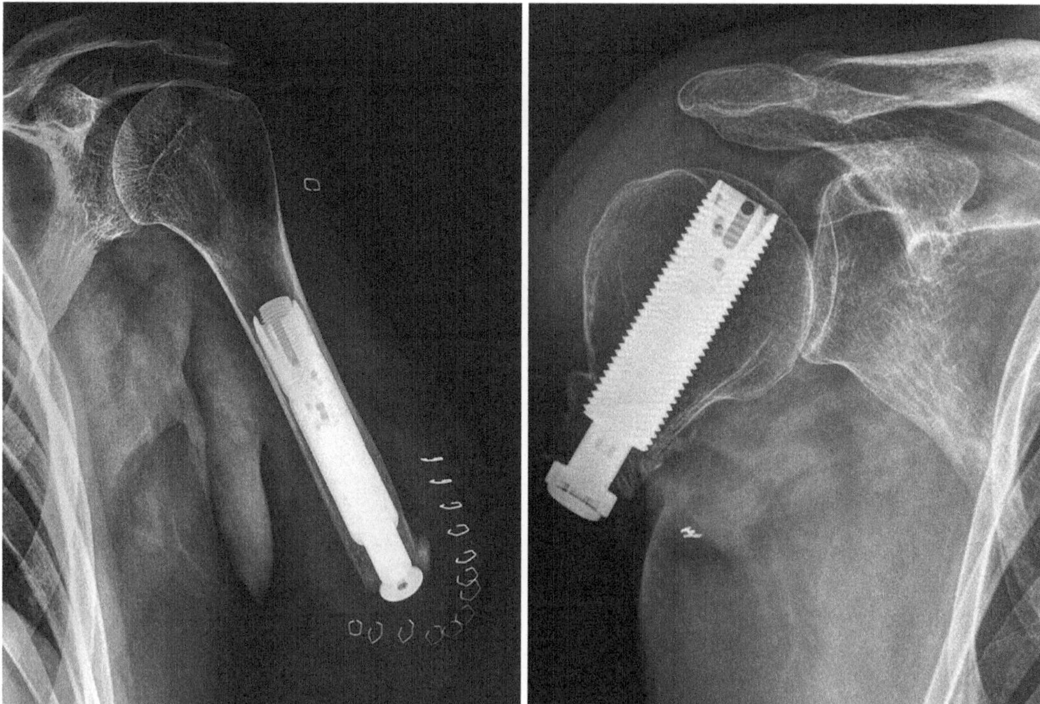

Figs. 4.9 and 4.10 X-ray after stage 1 osseointegration in a long and short transhumeral stump. The short stump has been augmented with a bone graft from the iliac crest to assure enough length for the implant

4.1.10 Secondary Indications

Usually, an amputation occurs due to trauma or oncologic surgery [30]. Next to these causes, certain critical injuries of the musculoskeletal system prohibit any functional ability of the hand. For example, global brachial plexus lesions and complex regional pain syndrome (CRPS) are severely disabling conditions, which leave the patient with an unrecoverable hand. Long-term consequences are social exclusion and psychologic disorders. Patients often report of a great burden living with the hypo- or hypersensitive and dysfunctional limb. Thus, in certain severe cases, a patient might be evaluated for an elective amputation, which pursues the following goals: reduction of pain, functional recovery, and enhancing quality of life [3, 4, 31].

A precondition is the exhaustion of biologic reconstruction methods, like nerve grafting, free functional muscle transfers, tendon transfers, and others. Usually, patients have a long medical history and accordingly alterations of the central representation of the affected hand as a functional body part. Prior to amputation, a semi-structured psychologic interview should evaluate the patient's qualification for the procedure, as well as individual motivations and expectations [32]. Only after a thorough evaluation and interdisciplinary discussion between the patient and the physician, psychologist, physiotherapist, and orthopedic technician is the patient eligible for an elective amputation. The amputation level is usually transradial, when shoulder and elbow function is present or has successfully been reanimated. However, delayed consultation after global plexopathies can also be an indication for transhumeral prosthetic reconstruction in combination with TMR.

Concerning the fitting of a prosthetic hand, a proper reconstructive strategy has to be chosen. A preoperative EMG assessment should detect potential signals for prosthetic control. After stable EMG signals have been acquired, the fitting of a hybrid prosthesis forms a proof of concept for the future prosthetic usage.

In cases, where nerval structures are damaged, the muscles at the stump region might be atrophied and dysfunctional. The surgical transplantation of a free functional muscle transplant can be performed to enhance the signal quality or quantity of the stump. The gracilis muscle serves as a reliable donor muscle, with a straightforward anatomy and long nerval trunk. Hence, it can be transplanted to the stump and reconnected to a functional proximal nerve end.

4.1.11 Rehabilitation

The focus of rehabilitation is to help the patient regain independence in everyday life as well as to return to occupation within a reasonable period, since unemployment and inactivity negatively affect patients' well-being and socioeconomic situation.

Wound healing and stump consolidation should be completed before prosthetic fitting is started. In the first phase, regular bandage and if necessary lymphatic drainage should be applied to prevent excessive stump swelling. In the upper extremity, first socket adaptions can usually be performed after 2–4 weeks; in the lower extremity, it might take a bit longer until a stable stump has formed. In cases of humeral osteosynthesis, weight loading should not be initiated prior to proper callus formation. For osseointegrative prostheses, gradual weight loading and axial pressure are applied according to the respective protocol, before the socket can be attached.

If applicable, specific training sessions with the physio- and occupational therapist should start even before amputation. The main goal is to detect and strengthen myoelectric signals by practicing with surface EMG [33]. The patient should have selective control over the present signals to assure dexterous prosthetic handling. In cases of selective nerve transfers, regeneration time will take at least 3 months until initial signals can be expected. Specific EMG training helps to improve volitional control of the reinnervated muscles. If a prolonged nerve regeneration time is expected, functional electrical stimulation using exponential current may be applied until the target muscles are reinnervated. Regular training with portable EMG devices and associated video games can significantly improve patient engagement and overall prosthetic performance [34, 35].

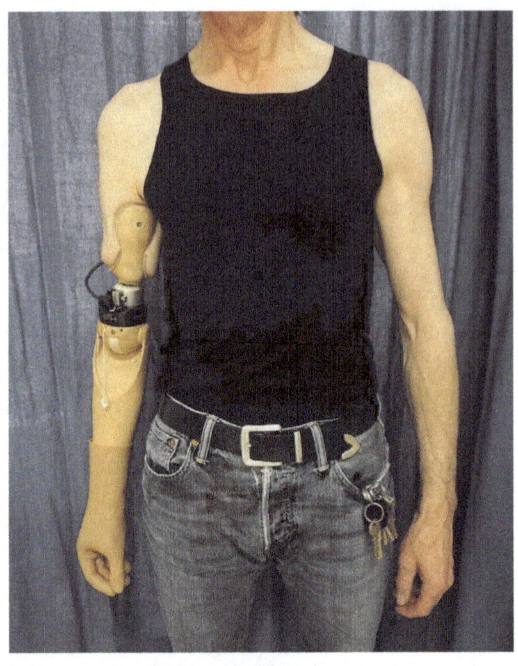

Fig. 4.11 Clinical appearance of a patient fitted with a transhumeral osseointegrated prosthesis with functional elbow joint. The superficial electrodes are placed on the ventral and dorsal side of the stump

Besides prosthetic function, physiotherapy should also focus on strengthening core muscles and balance, since amputees tend to adopt unphysiological posture (Fig. 4.11).

After the patient has successfully been equipped with the prosthetic device and routine usage is possible, follow-up consultations can be held on an individual basis. The most frequent issue in long-term follow-ups is stump or phantom limb pain. In the upper extremity, the soft tissue is only rarely compromised through prosthetic usage, which is more frequent in the lower extremity and might necessitate small surgical corrections. In osseointegrative patients, there is always a small risk for implant infection and implant loosening, which may be handled with either iterative revision surgery or extracorporeal shockwave therapy [36].

Nowadays, there is a wide range of options to surgically enable and improve the experience with an upper limb prosthesis. It involves all varieties of musculoskeletal surgery and individual concepts that are tailored to the single patient. Recent developments in nerve surgery have taken advantage of the great potential of biologic reconstruction. Still, the upcoming implantable technologies will have a great impact on patient treatment in the future and might pave the way towards a new standard of limb augmentation.

4.2 Hand Transplantation

By Frédéric Schuind, MD, PhD, from Brussels, Belgium.

In 1998, the world learned with stupefaction that a cadaveric hand had been successfully transplanted to a patient in France [37, 38]. One month later, a second hand was transplanted in the United States, and soon bilateral hand transplantations were performed. Actually, these were not the first attempts at transplanting a hand; a fist case had been done in South America in the 1960s, but, without modern immunosuppression, this first hand was rejected after 2 weeks only [39]. As usual in the case of a scientific advance raising ethical questions, the world was split into enthusiastic persons anticipating the development of a new area of medicine, where the loss of a human part could be perfectly replaced, including the foot, the face, the brain, or even the head, and opponents, representing the majority, believing that it was unethical to expose a patient to the risks of immunosuppression to regain the function of his/her hand: transplantation in their views had to be reserved to saving life. At that time however, renal transplantation did not raise concerns, while not saving the life of the recipient, but improving his/her quality of life, avoiding the burden of dialysis sessions. After learning about hand transplantation, John Irving wrote his famous novel "the Fourth Hand." The idea of transplantation of a cadaveric limb had actually a very long mythic history with the patrons of surgery, Cosmas and Damian, credited to have amputated a cancerous leg and replaced it with a leg from a Moor—many paintings celebrate this historical event.

The first cases of hand transplantation of the "modern" era demonstrated that such surgery is feasible, both surgically and immunologically, under immunosuppression comparable to the one prescribed for organ transplantation. Rejection episodes occur but can be readily detected by skin inspection and biopsy and reversed by appropriate adjustment of the medical treatment [40]. Hand/upper extremity transplantation is now considered among the group of vascularized composite allotransplantations (VCAs), comprising skin, subcutaneous, neurovascular, and mesenchymal tissues such as bone, cartilage, muscle, fascia, and skin [41]. The potential benefits of VCAs for defects caused by a trauma or resection for a tumor or related to a congenital deficiency are immense: complex reconstructions, presently performed using local or free flaps, never perfectly replicating the lost parts, could be accomplished by transplantation of matched tissues from a brain-dead donor, without donor-site morbidity. Other examples of successful clinical VCAs are face, larynx, abdominal wall, uterus, and lower extremity [39, 42–45]. All VCAs need immunosuppression to prevent rejection, especially targeted towards the transplanted skin, probably because of the existence in the epidermis of specific antigens. Because of the need of immunosuppression, because of the cost of this treatment, and because of the lack of donors, there have been less than 200 hand transplantations performed in the world since 1998 [46, 47]. The author of this chapter had the privilege to lead at his institution the team of a successful hand transplantation, performed in 2002 (Fig. 4.12). Almost 20 years later, the patient has still his transplanted hand, with acceptable function and good sensibility, though severe episodes of rejection have affected its cosmetic appearance (skin and nails) and caused fibrosis, progressively restricting finger flexion and strength—a form of chronic rejection [48, 49]. After this first case, a program of hand transplantation was instituted at the author's academic center, but no other case could be operated, because of a lack of donors and a

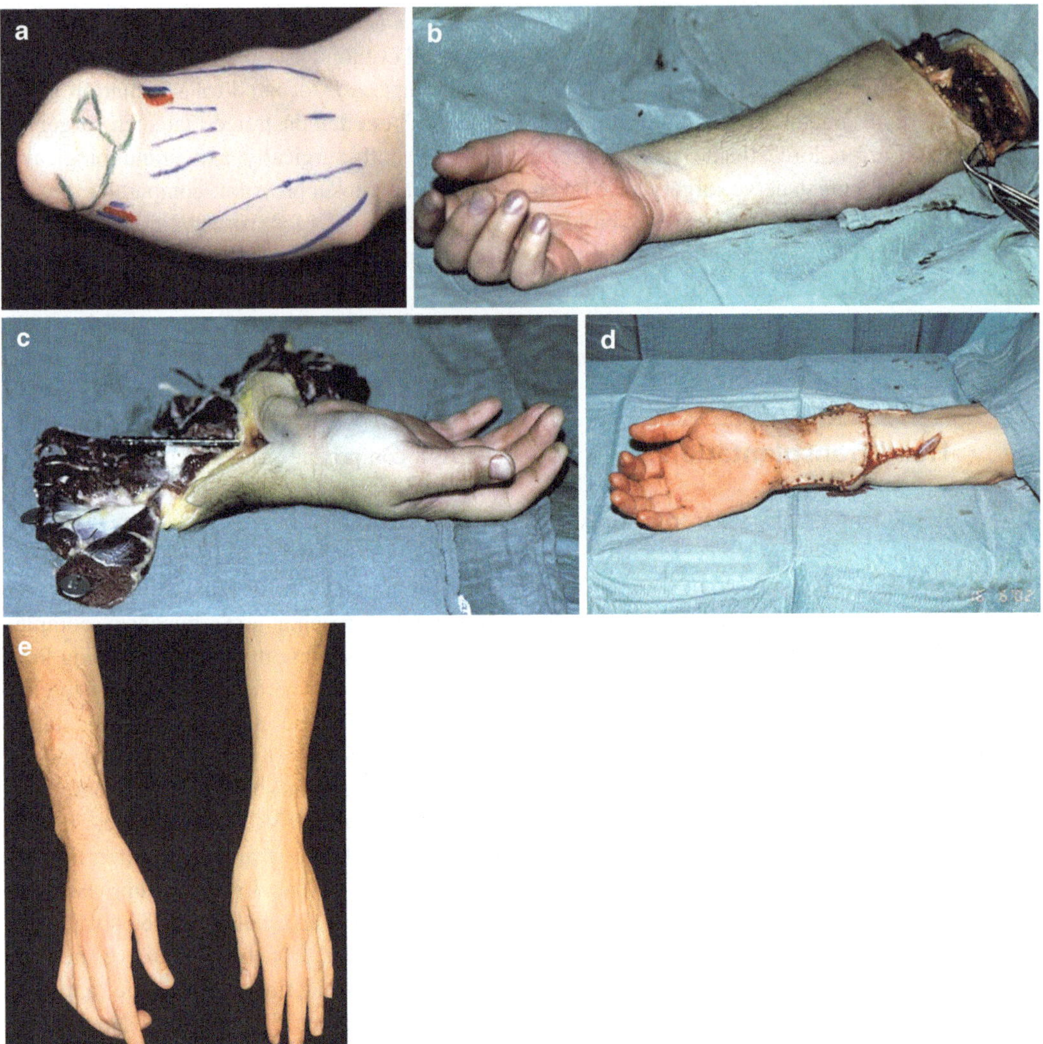

Fig. 4.12 Brussels hand transplantation case. (Reproduced with permission from Schuind F, Van Holder C, Mouraux D, Robert Ch, Meyer A, Salvia P, Vermeylen N, Abramowicz D. The first Belgian hand transplantation—37 month term results. J Hand Surg Br. 2006;31:371–376). (**a**) Preoperative forearm stump (note the drawing of superficial veins and deep arteries, and the presence of Tinel's sign at the level of the amputation neuromas). (**b–d**) Hand transplantation, surgical steps. (**b**) Elbow disarticulation. (**c**) The hand has been prepared for the transplantation; note the isolated musculotendinous units and the osteosynthesis plates already screwed to the radius and ulna. (**d**) Aspect of the hand at the end of the operation. (**e**) Aspects of the hands, 72 months after the hand transplantation, before the first rejection episode

local organization—despite the efforts of the hand surgeons and the existence of some good candidates for transplantation on the list and of a strong local research program on experimental composite allotransplantation [50, 51]. Based on this single case rewarding experience, disappointing for the clinical program of hand transplantation, the author provides in this chapter general recommendations to teams wishing to set up a program of hand transplantation at their institution. Specific details and precise guidelines can be found in the author's publication "Hand Transplantation: The State-of-the-Art" [40].

4.2.1 First, Constitute a Good Medico-Surgical Team

A successful hand transplant program is organized by an interdisciplinary consortium comprising (1) hand surgeons experienced in microsurgery and replantation, (2) a medico-surgical team active in transplantation of various organs, (3) physiotherapists and occupational therapists specialized in hand rehabilitation, (4) pathologists familiar with skin histology and organ transplantation, and (5) psychologists able to select the candidates to transplantation and to manage postoperative stress. The entire crew should meet on a regular basis, discussing and organizing the program of hand transplantation and solving local ethical issues. A detailed protocol should be established to manage the patients and adapted to the particular situations of each candidate recipient. Ideally, in parallel, a program of experimental research in limb transplantation should be organized as many unsolved questions still need to be answered regarding VCAs.

4.2.2 Second, Find a Good Candidate for Hand Transplantation

The ideal candidate is a highly motivated patient presenting the sequel of a "clean-cut" bilateral mid- or distal-forearm amputation. In this situation, the extrinsic muscles are preserved, allowing early postoperative active mobilization of the wrist and fingers; the vessels and nerves are intact almost up to the stump, allowing perfect anastomoses without tension. Less good candidates are those presenting a unilateral hand amputation, as the transplanted hand will never match the function and sensibility of the contralateral intact hand, though patients like the Brussels case with a unilateral hand transplantation have been quite satisfied of their operation. High-forearm and above-elbow amputations are other less good indications, because the hand function will depend entirely on nerve regeneration, though again acceptable results have been obtained after such transplantations [52]. Given the risks of long-term immunosuppression and the alternate possibilities of reconstruction, isolated finger or thumb transplantation and transplantation of finger segments, which are possible and indeed have been performed occasionally, are not considered reasonable, except maybe in a patient already under immunosuppression, for example after a previous kidney transplantation. In children, there is the question of the patient's consent. Indeed, limb transplantation could represent an answer to severe congenital deformities, but there are many uncertainties regarding future growth and long-term results. To the author's knowledge, such a program has not been developed yet.

The patient's strong wish and psychological stability to go through the whole transplantation process, with prolonged revalidation and lifelong immunosuppression, are one of the most important factors to consider in selecting the recipient candidate. Personality disorders are common among patients after a severe hand trauma [53–55], and many amputees have unrealistic expectations. The candidate should be perfect and several times informed of the burden and risks of chronic immunosuppression and of the anticipated incomplete results. Written consent does not dispense the team to provide continuous and repeated information and to answer all patient's questions. The local ethical committee might be consulted in special indications, for example if considering a transplantation in a child.

Other aspects are equally important. The candidate should be in good health. Preexistent oncologic and infectious diseases constituting absolute contraindications; abnormal renal, pulmonary, or cardiac function and relative contraindications; and whether the patient is or not seronegative for cytomegalovirus and Epstein-Barr virus are also important to consider. Medico-social aspects also need to be clarified. The costs of immunosuppression are enormous, and the social security/insurance companies do not routinely cover these expenses for a treatment still frequently considered experimental.

After this selection process, the long list of candidates for hand transplantation will be shrunk to a few or a single good recipient candidate(s), who will probably need to wait a long time before the team finds a suitable donor.

4.2.2.1 You Also Need to Find a Good Donor

The donor should be matched with the recipient for blood group, size, gender, and cosmetic appearance. In case of a trauma, the upper extremity must be intact from fracture or wound. A good HLA matching is desirable but rarely possible. According to your national legislation, consent by the donor's family might be an obligation—in any case, the family will be informed about the hand transplantation and of course notice the missing hand replaced by a cosmetic prosthesis, so a discussion between family members and the transplantation coordinator always needs to take place. It is our experience that many times the family agrees for the harvesting of organs, but not for the hand(s). A preoperative lymphocytotoxic crossmatch between donor and recipient must be obtained and be negative. Potential donors will therefore also shrink to very few, much less than the number of donors for general transplantation.

4.2.3 Surgical Considerations

Hand transplantation has common points and differences with hand replantation. Both are emergency operations. Ischemia time critically influences muscle function and nerve regeneration, so rapid revascularization after stable bone fixation is mandatory to achieve later on acceptable function and sensibility. Prolonged ischemia might also have immunological consequences favoring chronic rejection and has less clear effects on acute rejection and cytomegalovirus replication. Ischemia time is difficult to control, especially if the donor is in a hospital distant to the transplant center and if precedence is given to other organs' harvesting. We recommend if possible to operate donor and recipient in adjacent theatres of the same institution and to harvest very quickly, under tourniquet that may remain until final management of the cadaveric body, the hand(s), before the solid organs—this needs to be agreed in advance with the other transplantation surgeons. Another important point to manage in advance is the repartition of the surgical, anesthesiologist, and nursing staff in the operation rooms as several transplantations will probably be performed simultaneously from one single good-quality donor.

After perfusion of a preservation solution, the hand(s) is(are) prepared on a side table, removing the excess tissues and shortening bones, nerves, vessels, and tendons to precisely match the recipient's forearm(s) and to allow nerve and vascular sutures without any tension. At the same time, a second team prepares the recipient's stump(s). The transplantation procedure is then similar to replantation, starting by osteosynthesis, followed by revascularization, then tendon and nerve repair, and finally skin suture.

4.2.4 Medical Considerations During and After the Transplantation

The operation is done under antibioprophylaxis, including against cytomegalovirus and *Pneumocystis carinii*. Aspirin or low-molecular subcutaneous heparin could prevent postoperative vascular thrombosis. Immunosuppression is started at the induction of the operation, following a precise protocol decided in advance, taking into account the specificities of hand transplantation: hand transplantation is not lifesaving (contrarily for example to heart transplantation); hand-transplanted patients are in principle healthy and lack comorbidities; the diagnosis of rejection is easy as it primarily affects the skin of the transplanted hand, constantly visible. Many immunosuppression protocols are available, but reporting these is beyond the scope of this chapter. In any case and despite the efforts to find an alternative ("tolerance induction"), lifelong immunosuppression remains indispensable. Acute rejection episodes have occurred in almost all hand-transplant recipients. Acute rejection not only represents an imminent danger to graft survival but also has an unknown impact on the development of chronic rejection with progressive loss of the function of the hand.

Immunosuppression exposes the patient to various risks including opportunistic infections, malignancy, and metabolic complications like diabetes mellitus, hypertension, kidney dysfunction, and osteonecrosis [56, 57]. Graft-versus-host disease has been feared after hand transplantation but to the author's knowledge never reported. Currently, protocols for minimization of immunosuppression are recommended, which could reduce the rate of complications, at the possible price of quicker and more frequent phenomena of chronic rejection.

4.2.5 Rehabilitation

Prolonged physiotherapy is essential to the functional success after hand transplantation. The program includes reeducation of sensibility and global motion pattern, cortical reprogramming, and, later, occupational therapy.

4.2.6 Conclusions

The author is convinced of the place of VCAs in the treatment of selected—actually rare—cases of limb amputation. The most common form of a VCA is hand transplantation. The outcome after such surgery is critically dependent on the patient's psychological status, compliance, and motivation, as well as on the efforts of the surgical and rehabilitation team. The clinical result is frequently better than expected—better than after an equivalent replantation. The satisfaction of the patient is high, as, in addition to recovering hand function, he/she enjoys the restoration of body integrity, which is to him/her essential and difficult to quantify. The whole process can be quite rewarding to the medico-surgical team. However, the team must be prepared to severe complications that may occur, sometimes years after the transplantation, imposing on occasion the arrest of the immunosuppression leading usually to reamputation. Setting up a program of hand transplantation remains a challenge, and this type of surgery should be reserved to academic institutions with highly motivated teams and supportive directions.

References

1. Ziegler-Graham K, MacKenzie EJ, Ephraim PL, Travison TG, Brookmeyer R. Estimating the prevalence of limb loss in the United States: 2005 to 2050. Arch Phys Med Rehabil. 2008;89:422–9.
2. Roche AD, et al. A structured rehabilitation protocol for improved multifunctional prosthetic control: a case study. J Vis Exp. 2015;(105):e52968.
3. Aszmann OC, et al. Elective amputation and bionic substitution restore functional hand use after critical soft tissue injuries. Sci Rep. 2016;6:34960.
4. Aszmann OC, et al. Bionic reconstruction to restore hand function after brachial plexus injury: a case series of three patients. Lancet. 2015;385:2183–9.
5. Pierrie SN, Gaston RG, Loeffler BJ. Current concepts in upper-extremity amputation. J Hand Surg. 2018;43:657–67.
6. Fitzgibbons P, Medvedev G. Functional and clinical outcomes of upper extremity amputation. J Am Acad Orthop Surg. 2015;23:751–60.
7. Morgan EN, Potter BK, Souza JM, Tintle SM, Nanos GP. Targeted muscle reinnervation for transradial amputation: description of operative technique. Tech Hand Upper Extrem Surg. 2016;20:166–71.
8. Kusnezov N, Dunn JC, Stewart J, Mitchell JS, Pirela-Cruz M. Acute limb shortening for major near and complete upper extremity amputations with associated neurovascular injury: a review of the literature. Orthop Surg. 2015;7:306–16.
9. Resnik L, Borgia M, Biester S, Clark MA. Longitudinal study of prosthesis use in veterans with upper limb amputation. Prosthetics Orthot Int. 2020;45(1):26–35. https://doi.org/10.1177/0309364620957920.
10. Wright TW, Hagen AD, Wood MB. Prosthetic usage in major upper extremity amputations. J Hand Surg. 1995;20:619–22.
11. Tsikandylakis G, Berlin Ö, Brånemark R. Implant survival, adverse events, and bone remodeling of osseointegrated percutaneous implants for transhumeral amputees. Clin Orthop Relat Res. 2014;472:2947–56.
12. Aszmann OC, Rab M, Kamolz L, Frey M. The anatomy of the pectoral nerves and their significance in brachial plexus reconstruction. J Hand Surg. 2000;25:942–7.
13. Tukiainen E, Barner-Rasmussen I, Popov P, Kaarela O. Forequarter amputation and reconstructive options. Ann Plast Surg. 2020;84:651–6.
14. Kuiken TA, Dumanian GA, Lipschutz RD, Miller LA, Stubblefield KA. The use of targeted muscle reinnervation for improved myoelectric prosthesis control in a bilateral shoulder disarticulation amputee. Prosthetics Orthot Int. 2004;28(3):245–53.
15. Hoffer JA, Loeb GE. Implantable electrical and mechanical interfaces with nerve and muscle. Ann Biomed Eng. 1980;8(4–6):351–60.
16. Kuiken TA, Barlow AK, Hargrove LJ, Dumanian GA. Targeted muscle reinnervation for the upper and lower extremity. Tech Orthop. 2017;32:109–16.

17. Gart MS, Souza JM, Dumanian GA. Targeted muscle reinnervation in the upper extremity amputee: a technical roadmap. J Hand Surg. 2015;40:1877–88.
18. Salminger S, et al. Outcomes, challenges, and pitfalls after targeted muscle reinnervation in high-level amputees: is it worth the effort? Plast Reconstr Surg. 2019;144:1037e–43e.
19. Poppler LH, et al. Surgical interventions for the treatment of painful neuroma: a comparative meta-analysis. Pain. 2018;159:214–23.
20. Santosa KB, Oliver JD, Cederna PS, Kung TA. Regenerative peripheral nerve interfaces for prevention and management of neuromas. Clin Plast Surg. 2020;47:311–21.
21. Souza JM, et al. Targeted muscle reinnervation: a novel approach to postamputation neuroma pain. Clin Orthop Relat Res. 2014;472:2984–90.
22. Lanier ST, Jordan SW, Ko JH, Dumanian GA. Targeted muscle reinnervation as a solution for nerve pain. Plast Reconstr Surg. 2020;146:651E–63E. https://doi.org/10.1097/PRS.0000000000007235.
23. Dumanian GA, et al. Targeted muscle reinnervation treats neuroma and phantom pain in major limb amputees: a randomized clinical trial. Ann Surg. 2019;270:238–46.
24. Urbanchek MG, et al. Development of a regenerative peripheral nerve interface for control of a neuroprosthetic limb. Biomed Res Int. 2016;2016:5726730.
25. Vu PP, et al. A regenerative peripheral nerve interface allows real-time control of an artificial hand in upper limb amputees. Sci Transl Med. 2020;12:1–12.
26. Clites TR, Herr HM, Srinivasan SS, Zorzos AN, Carty MJ. The Ewing amputation: the first human implementation of the agonist-antagonist myoneural interface. Plast Reconstr Surg Glob Open. 2018;6:e1997.
27. Clites TR, Carty MJ, Srinivasan S, Zorzos AN, Herr HM. A murine model of a novel surgical architecture for proprioceptive muscle feedback and its potential application to control of advanced limb prostheses. J Neural Eng. 2017;14:036002.
28. Clemente F, et al. Touch and hearing mediate osseoperception. Sci Rep. 2017;7:45363.
29. Ortiz-Catalan M, Mastinu E, Sassu P, Aszmann O, Brånemark R. Self-contained neuromusculoskeletal arm prostheses. N Engl J Med. 2020;382:1732–8.
30. Tintle SM, Baechler MF, Nanos GP, Forsberg JA, Potter BK. Traumatic and trauma-related amputations: Part II: Upper extremity and future directions. J Bone Joint Surg A. 2010;92:2934–45.
31. Hruby LA, et al. Bionic upper limb reconstruction: a valuable alternative in global brachial plexus avulsion injuries—a case series. J Clin Med. 2019;9:23.
32. Hruby LA, Pittermann A, Sturma A, Aszmann OC. The Vienna psychosocial assessment procedure for bionic reconstruction in patients with global brachial plexus injuries. PLoS One. 2018;13:e0189592.
33. Sturma A, Hruby LA, Prahm C, Mayer JA, Aszmann OC. Rehabilitation of upper extremity nerve injuries using surface EMG biofeedback: protocols for clinical application. Front Neurosci. 2018;12:906.
34. Prahm C, Vujaklija I, Kayali F, Purgathofer P, Aszmann OC. Game-based rehabilitation for myoelectric prosthesis control. JMIR Serious Games. 2017;5:e3.
35. Prahm C, Kayali F, Sturma A, Aszmann O. PlayBionic: game-based interventions to encourage patient engagement and performance in prosthetic motor rehabilitation. PM R. 2018;10:1252–60.
36. Gstoettner C, et al. Successful salvage via re-osseointegration of a loosened implant in a patient with transtibial amputation. Prosthetics Orthot Int. 2020;45(1):76–80. https://doi.org/10.1177/0309364620953985.
37. Dubernard JM, Owen E, Herzberg G, Martin X, Guigal V, Dawahra M, Pasticier G, Mongin-Long D, Kopp C, Ostapetz A, Lanzetta M, Kapila H, Hakim N. Première transplantation de main chez l'homme. Résultats précoces. Chirurgie. 1999;124:358–65. https://doi.org/10.1016/s0001-4001(00)80007-0.
38. Dubernard JM, Owen E, Herzberg G, Lanzetta M, Martin X, Kapila H, Dawahra M, Hakim NS. Human hand allograft: report on first 6 months. Lancet. 1999;353(9161):1315–20. https://doi.org/10.1016/S0140-6736(99)02062-0.
39. Barker JH, Francois CG, Frank JM, Maldonado C. Composite tissue allotransplantation. Transplantation. 2002;73:832–5. https://doi.org/10.1097/00007890-200203150-00033.
40. Schuind F, Abramowicz D, Schneeberger S. Hand transplantation: the state-of-the-art. J Hand Surg Eur. 2007;32:2–17. https://doi.org/10.1016/j.jhsb.2006.09.008.
41. Murphy BD, Zuker RM, Borschel GH. Vascularized composite allotransplantation: an update on medical and surgical progress and remaining challenges. J Plast Reconstr Aesthet Surg. 2013;66:1449–55. https://doi.org/10.1016/j.bjps.2013.06.037.
42. Carty MJ, Zuker R, Cavadas P, Pribaz JJ, Talbot SG, Pomahac B. The case for lower extremity allotransplantation. Plast Reconstr Surg. 2013;131:1272–7. https://doi.org/10.1097/PRS.0b013e31828bd1a5.
43. Hofmann GO, Kirschner MH. Clinical experience in allogeneic vascularized bone and joint allografting. Microsurgery. 2000;20:375–83. https://doi.org/10.1002/1098-2752(2000)20:8<375::aid-micr6>3.0.co;2-0.
44. Levi DM, Tzakis AG, Kato T, Madariaga J, Mittal NK, Nery J, Nishida S, Ruiz P. Transplantation of the abdominal wall. Lancet. 2003;361(9376):2173–6. https://doi.org/10.1016/S0140-6736(03)13769-5.
45. Siemionow M. Vascularized composite allotransplantation: a new concept in musculoskeletal regeneration. J Mater Sci Mater Med. 2015;26:266. https://doi.org/10.1007/s10856-015-5601-5.
46. Petruzzo P, Dubernard JM. The international registry on hand and composite tissue allotransplantation. Clin Transpl. 2011:247–53.
47. Shores JT, Brandacher G, Andrew Lee WP. Hand and upper extremity transplantation: an update of outcomes in the worldwide experience. Plast Reconstr

Surg. 2015;135:351e–60e. https://doi.org/10.1097/PRS.0000000000000892.
48. Schneeberger S, Gorantla VS, van Riet RP, Lanzetta M, Vereecken P, van Holder C, Rorive S, Remmelink M, Le Moine A, Abramowicz D, Zelger B, Kaufman CL, Breidenbach WC, Margreiter R, Schuind F. Atypical acute rejection after hand transplantation. Am J Transplant. 2008;8:688–96. https://doi.org/10.1111/j.1600-6143.2007.02105.x.
49. Schuind F, Van Holder C, Mouraux D, Robert C, Meyer A, Salvia P, Vermeylen N, Abramowicz D. The first Belgian hand transplantation—37 month term results. J Hand Surg Br. 2006;31:371–6. https://doi.org/10.1016/j.jhsb.2006.01.003.
50. Li Z, Benghiat FS, Kubjak C, Schuind F, Goldman M, Le Moine A. Donor T-cell development in host thymus after heterotopic limb transplantation in mice. Transplantation. 2007;83:815–8. https://doi.org/10.1097/01.tp.0000255703.02587.df.
51. Li Z, Benghiat FS, Charbonnier LM, Kubjak C, Rivas MN, Cobbold SP, Waldmann H, De Wilde V, Petein M, Schuind F, Goldman M, Le Moine A. CD8+ T-Cell depletion and rapamycin synergize with combined coreceptor/stimulation blockade to induce robust limb allograft tolerance in mice. Am J Transplant. 2008;8:2527–36. https://doi.org/10.1111/j.1600-6143.2008.02419.x.
52. Iglesias M, Ramírez-Berumen M, Butrón P, Alberú-Gómez J, Salazar-Hernández F, Macias-Gallardo J, Leal-Villalpando RP, Zamudio-Bautista J, Acosta V, Jauregui L, Hernández-Campos A, Espinosa-Cruz V, Vázquez-Lamadrid J, González-Sánchez J, Cuellar-Rodriguez J, Sierra-Madero JG, Gaytan-Cervantes R, Contreras-Barbosa S, Navarro-Lara A, Guzman-Gonzalez J, Domínguez-Cherit J, Vilatoba M, Toussaint-Caire S, Vega-Boada F, Gómez-Pérez FJ, Mayorquin-Ruiz M. Functional outcomes 18 months after total and midarm transplantation: a case report. Transplant Proc. 2018;50:950–8. https://doi.org/10.1016/j.transproceed.2017.12.027.
53. Meyer TM. Psychological aspects of mutilating hand injuries. Hand Clin. 2003;19:41–9. https://doi.org/10.1016/s0749-0712(02)00056-2.
54. Schweitzer I, Rosenbaum MB, Sharzer LA, Strauch B. Psychological reactions and processes following replantation surgery: a study of 50 patients. Plast Reconstr Surg. 1985;76:97–103.
55. Sims AC. Psychogenic causes of physical symptoms, accidents and death. J Hand Surg Br. 1985;10:281–2. https://doi.org/10.1016/s0266-7681(85)80043-7.
56. Conrad A, Petruzzo P, Kanitakis K, Gazarian A, Badet L, Thaunat O, Vanhems P, Buron F, Morelon E, Sicard A. DIVAT consortium and the IRHCTT teams. Infections after upper extremity allotransplantation: a worldwide population cohort study, 1998-2017. Transpl Int. 2019;32:693–701. https://doi.org/10.1111/tri.13399.
57. Krezdorn N, Tasigiorgos S, Wo L, Lopdrup R, Turk M, Kiwanuka H, Ahmed S, Petruzzo P, Bueno E, Pomahac B, Riella LV. Kidney dysfunction after vascularized composite allotransplantation. Transplant Direct. 2018;4:e362. https://doi.org/10.1097/TXD.0000000000000795.

5 Specific Peripheral Nerve Surgery for the Upper Limb

J. Bahm, *Surgical Rationales in Functional Reconstructive Surgery of the Upper Extremity*,
https://doi.org/10.1007/978-3-031-32005-7_5

5.1 Knowledge on Peripheral Nerve De- and Regeneration

This complex biological healing process is actually well described in articles focusing on neuropathology and morphology [1, 2] and biophysiology (signaling, growth factors ...) (Figs. 5.1 and 5.2).

Proximal nerve root lesions are known to induce motoneuron apoptosis, and the speed of regeneration (always estimated at about 1 mm per day) is the major decision point for any motor reconstructive goal or rehabilitation protocol.

Concerning morphology, the regeneration cones need the basal membrane to proceed along into the good (distal) direction. Therefore, at the nerve coaptation site, extreme precision is mandatory to avoid axonal loss or misrouting.

Electrophysiologic measurements are reliable and valuable only several weeks after nerve injury, as the characteristic features of denervation need some time to install their morphologic and pathophysiologic background, which then may become detectable by electrophysiologic measurements.

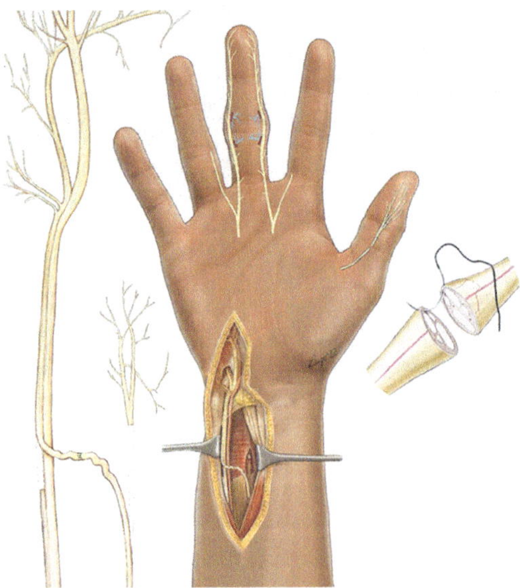

Fig. 5.1 The biological processes of de- and reinnervation and the surgical tools of nerve coaptation and nerve transfer to reestablish proper reinnervation.

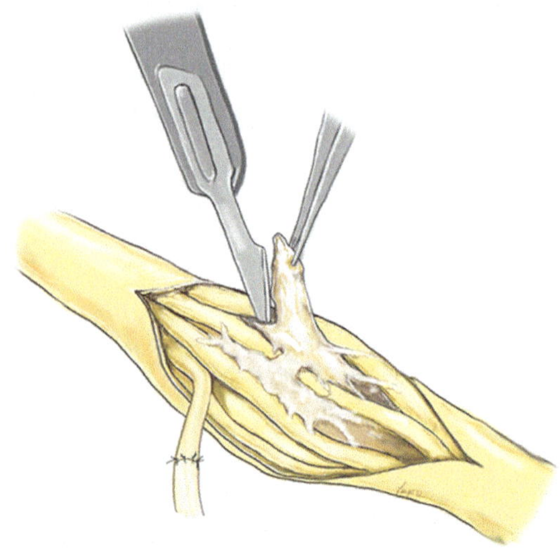

Fig. 5.2 Principles of nerve fascicle selection and selective nerve transfer technique.

Little is known about specific motor or sensory fiber regeneration; it seems that for nerve grafting, a defect in a pure motor nerve is better treated with grafts harvested from a motor nerve—something rather unpracticable in human surgery due to the new donor defect.

Also, our knowledge and understanding of inner topography of peripheral nerves with their plexiform architecture are rather limited, resulting in a certain source for axonal misdirection and limited function.

So far, there is no medication able to reliably boost the regeneration process, no growth factor to be added to our fibrin glue in nerve coaptation microsurgery, and no additional tool to fasten and enhance the regeneration process. First attempts with intraoperative direct electrostimulation of the nerve coaptation site are promising [3].

There is still debate if the regeneration process continues equally over longer distances (and longer time periods than 12–18 months) with the same power and efficiency. Especially, the contralateral nerve transfer surgery [4] and attempts to treat paraplegic patients by long grafts coaptated to nerve transfers from the upper limb down to target muscles at the proximal lower limb to restore stability for the standing position or elemental gait movements bear the problem of long-distance reinnervation potential.

Related experimentation in animals actually does not correlate to the human situation (limb size, time delay of several years impossible to handle in animal experimentation).

Although a lot of energy goes into research on peripheral nerve regeneration and related bioartificial nerve grafts, the knowledge about the primary targets (end organs), like the motor end plate or the different sensory organs, is still not sufficient. This lacks in our common clinical knowledge and daily decision-making.

Therefore, we add ideas and knowledge on the motor end plate (Sect. 5.2) and still wait for more interest concerning for example the mechanism and the behavior of the organs of sensation (neuroscience research fields).

5.2 Muscle Atrophy, the Motor End Plate

Denervation amyotrophy is a clinically well-known phenomenon, with its surgical correlate of wasting muscle mass, connective tissue, and fatty infiltration (both seen on ultrasound and MR imaging and surgical exploration), and time-dependent lack of functional recovery, even after proper nerve repair and regeneration.

In successfully reinnervated muscles, it remains a miracle to observe how the tensile strength regulation reappears, how the work between agonists and antagonists is re-established, and how the cortical control may reappropriate conscious and voluntarily controlled function.

The motor end plate seems to be one of the critical elements in long-term motor reinnervation.

Those end plates change their shape and structure with aging, and their behavior after nerve degeneration is complex and the information gained from animal experimentation huge and sometimes contradictory [5]. The motor end plate does not just disappear over time after unsuccessful reinnervation, but it does not "simply" recover after sound (surgical) reinnervation.

One interesting research protocol could be the examination of different settings of direct neuromuscular neurotization [6], where endings of motor nerves are directly transposed into viable muscle tissue, where they induce the formation of new motor end plates.

Also, viable motor end plates are not the end point, as the information translated from an electrical signal to the acetylcholine transmitter and its receptor induces another electrical signal in the muscle fiber, inducing fiber shortening. This path from the ACh receptor to the fiber is another interesting segment not much explored so far.

Electrical stimulation provided in due manner (wave form characteristics, frequency, time of application) by an appropriated external device reduces denervation amyotrophy (functional electrical stimulation) and is helpful in clinical settings as long as the regular muscle activity by endogenous innervation is interrupted. No other strategies are provided so far to reduce or delay amyotrophy or to act on the muscle components themselves—e.g., allowing a "maintenance" of the motor end plate until the arrival of newly regenerated motor axons.

5.3 Basic Microsurgical Techniques

The microanatomy of the peripheral nerve, like represented in Fig. 5.3, is a common base for this chapter. The surgical identification of a peripheral nerve itself in its healthy tissue bed cannot be assimilated to a neurolysis; only when there is a scarred bed due to trauma or a previous surgery can we define this exposure as an external neurolysis, where the nerve is freed from paraneural adherences or fibrosis, affecting its natural gliding capacities.

Millesi classified the fibrotic lesions affecting the different inner layers of the peripheral nerve (Fig. 5.4) and proposed the adequate surgical solutions, i.e., *internal neurolysis* after epineurotomy when the extrafascicular structures are involved (thereby defining different degrees of internal neurolysis) or *grafting* when the fascicles are too fibrotic themselves [7]. This

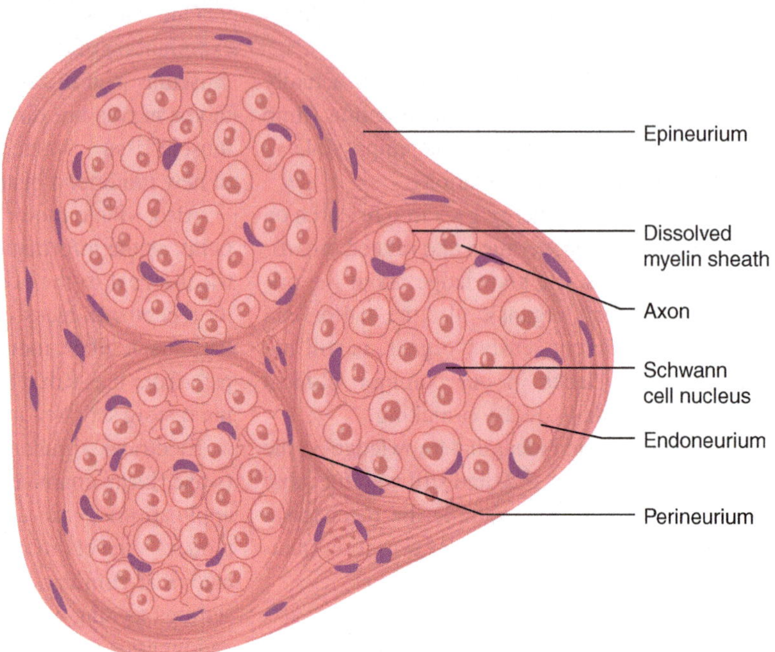

Fig. 5.3 Microanatomy of a peripheral nerve—paraneurium, epi-perineurium, fascicles, blood supply, and gliding tissue

morphologic evaluation may be completed with intraoperative direct electrical stimulation, which adds a functional argument (the so-called *conducting* neuroma).

Epineurotomy opens the inner nerve space and allows the postoperative development of interfascicular fibrosis with sometimes severe temporary or permanent worsening of nerve function. The epineurium comprises the tiny collateral vessels irrigating the nerve, and their trauma during epineurotomy or epineurectomy may severely affect the local nerve industry by ischemia (neurapraxia).

Epineurotomy is no longer a routine step in peripheral nerve decompression, like in carpal or ulnar tunnel surgery, or thoracic outlet, as this step has proven to add nothing positive to the functional outcome (guidelines CTS; [8]). But epineurotomy is a mandatory step in every nerve transfer surgery, as it allows access to a careful dissection of the fascicular groups in order to select redundant donor fascicle(s) or to choose the right recipient nerve parts—allowing a selective motor-to-motor or sensible-to-sensible nerve transfer, which actually seems to give the best outcome.

One negative aspect of Millesi's classification is that so far, the severe fibrosis of inner structures can only be identified after nerve transection, with observation of the cut slices, which will definitely suppress any remaining function and downgrade the nerve conduction capacity postoperatively. Ultrasound with high resolution may soon become the pre- and intraoperative tool of choice to overcome this difficulty (Fig. 5.5).

On the other hand, the ratio behind Millesi's description allows an approximation towards the pathophysiologic classification of Seddon and Sunderland, bringing together the more theoretical-conceptual analysis of the latter with the clinical intraoperative judgement.

Every microsurgical action on a peripheral nerve has to be performed under microsurgical conditions, i.e., using adequate magnification (magnifying glasses or operative microscope) and microsurgical instruments (forceps, scissors, needle holder). The basic requirements for microsurgery must be applied: a calm environment, stable operator's wrists and forearms, good light, and careful handling.

Nerve end coaptation is performed with thin nylon filament, by epineural or epi-perineural

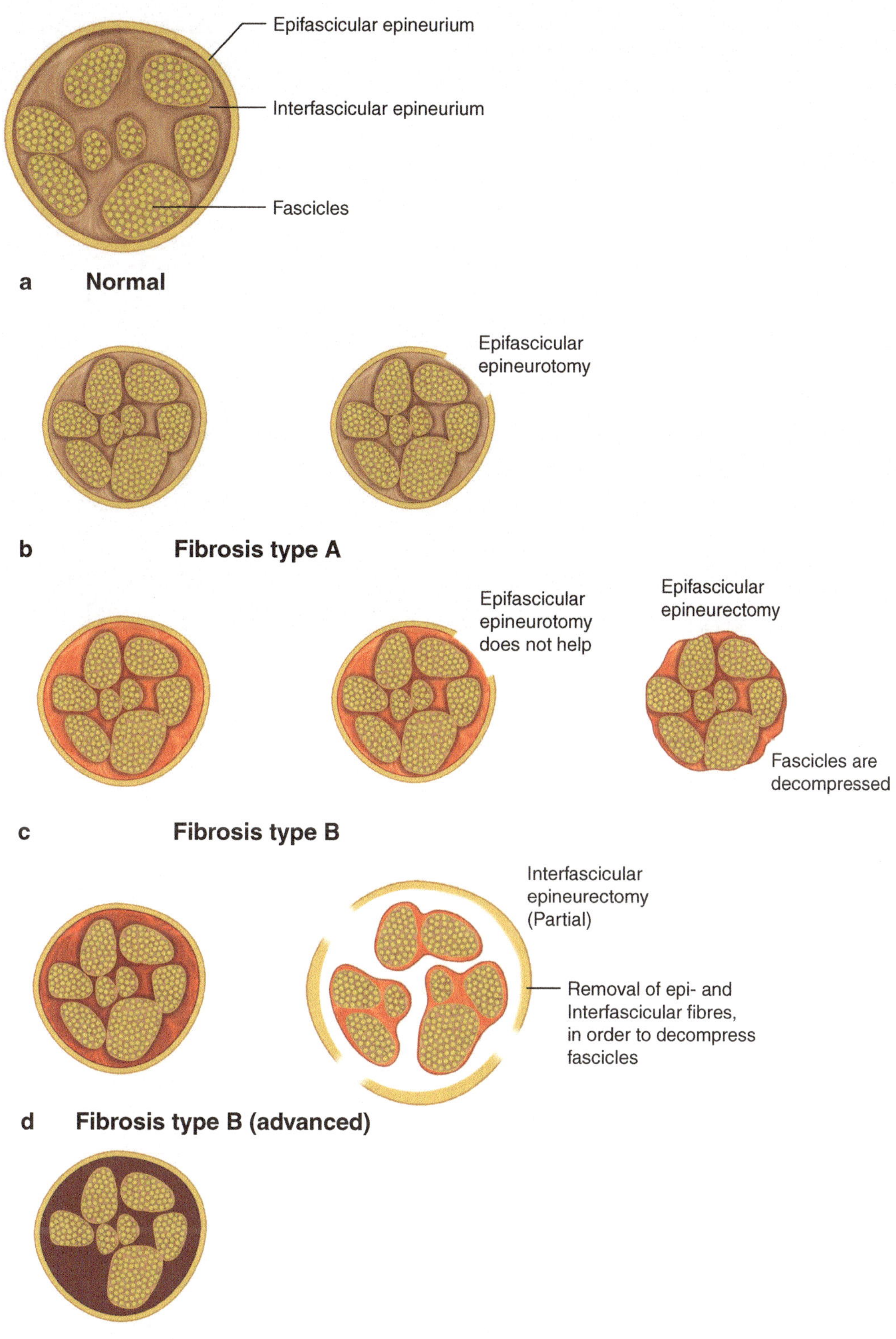

Fig. 5.4 Millesi classification: figure (**a**) shows a normal nerve, (**b**–**e**) different types of fibrotic tissue invasion

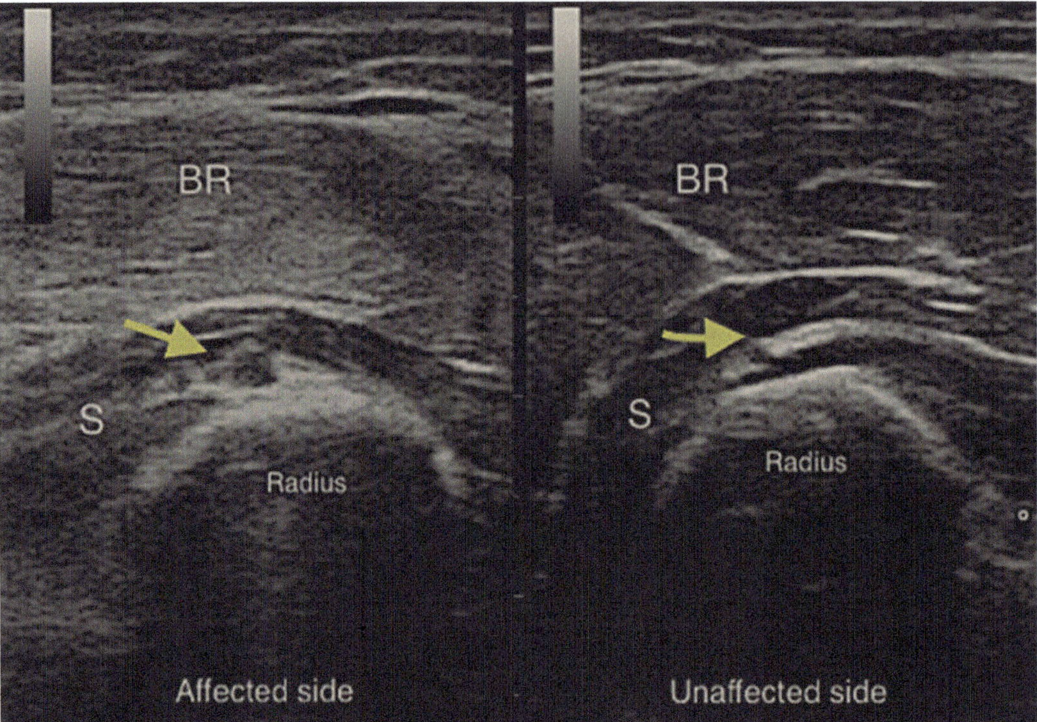

Fig. 5.5 Ultrasound image of peripheral nerve

stitches. In larger nerves, like brachial plexus trunks, the epineurium might be resistant alone; in smaller nerves, the needle bite mostly takes also the outer perineurial layer.

Coaptation may be safely assured by stitches or perineural fibrin glue application; in the absence of local tension, the results are of equal quality.

The general dogma of "absence of tension" remains rather badly defined, as there are few animal research works on this topic and also because little is known about the tolerable amount of local tension in smaller or bigger nerves. Obviously, limited gaps may be overcome by wider dissection and proximal mobilization of the distal nerve part, and a suture frame with 6-0 stitches which will decrease the local tension with every added stitch under good conditions (Fig. 5.6). It remains obvious that excessive tension is deleterious on coaptation sites for tiny nerves, but that adequate mobilization of stumps allows good-quality suture and functional recovery in bigger nerves, thereby avoiding short grafts with the greater

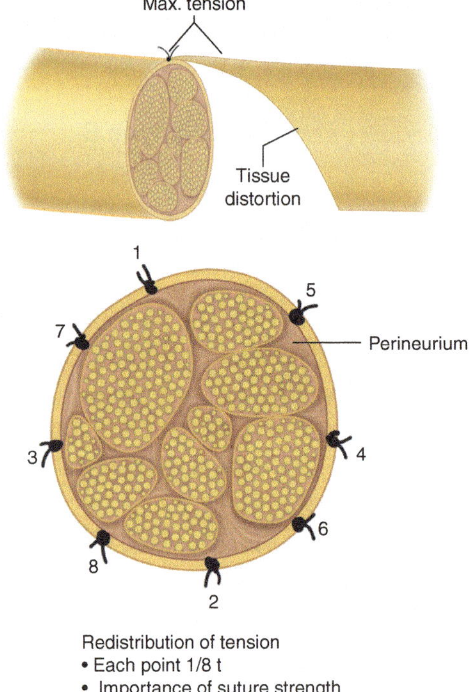

Fig. 5.6 Distribution of tension in a nerve coaptation suture

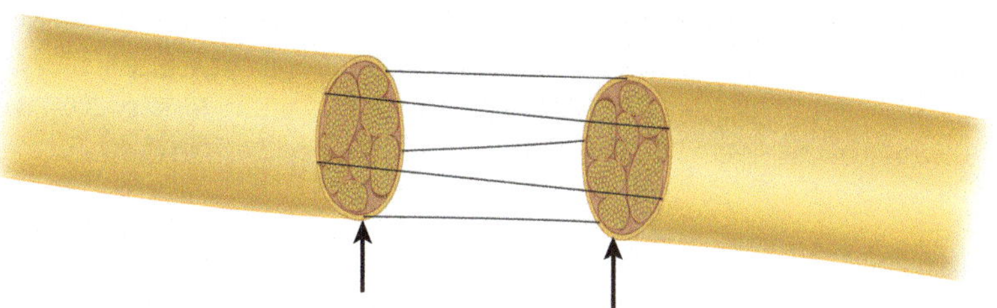

Fig. 5.7 Two coaptation sites with a cable graft

risk of fascicular mismatch and the necessity of two coaptation sites when using a graft (Fig. 5.7).

This attitude should not prevent the surgeon from avoiding tension on any coaptation site whenever possible, especially in nerve transfer surgery, where both nerve ends, donor and recipient, may be dissected out and cut with sufficient nerve length to avoid any tension (following the mantra: donor most distal, recipient most proximal).

5.4 Nerve Decompression

Decompressive surgery may be limited to the opening of compartment-like structures, like the carpal or Guyon tunnel, or the ulnar nerve sulcus. No real nerve surgery needs to be added in a primary decompressive surgery (there is no "neurolysis").

Thoracic outlet surgery is a particular decompression scenario, as multiple compression sites must be investigated and the presence of abnormal muscles, bands, and cervical ribs must be searched for and addressed properly, step-by-step.

If the nerve is altered by an intrinsic neuropathy, like the diabetic changes including microangiopathy, early decompression seems more beneficial.

5.4.1 No Epineurectomy, No Interfascicular Neurolysis

We have unfortunately observed patients operated elsewhere for a thoracic outlet syndrome by extensive neurolysis and large epineurotomies and interfascicular neurolysis, ending up with severe and progressive postoperative intraneural scarring and an extensive loss of their motor function, and also developing severe neuropathic pain. In the first observed case, I had the opportunity to return into the operative field, and had to perform a difficult and extended dissection of the soft tissues, to identify the nerve trunks which were totally fibrotic, hard on palpation, not conducting on direct electrical stimulation, and impossible to clean further to come back to a multifascicular structure. Not only did I have to stop at this level of dissection, as the only alternative was a complete segmental resection and interposition grafting, but also was there no more healthy gliding tissue locally available to conceive a regional adipo-fascial flap to cover the previously operated nerve structures, to protect them from new postoperative fibrosis and adherence. Only a free flap or a huge omental pedicled flap could have been helpful, and the patient did not accept further and more extensive surgery. In those cases, the functional prognosis seems to be very limited, in terms of recovery of motor function, but moreover considering chronic and severe neuropathic pain.

It has been shown for the surgery of carpal tunnel release that epineurotomy does not add anything beneficial for the nerve and the functional outcome. So why use this in more extended compression neuropathies, especially when the local environment is prone to postoperative adherences (very well known by brachial plexus surgeons who never return to an initial supraclavicular brachial plexus exposure)?

5.5 Nerve Reconstruction

Reconstruction of nerve continuity is assumed by direct suture, allowing fascicular groups to face without disturbing the inner architecture, and respecting the plexus structure of a peripheral

Table 5.1 Common nerve transfers in the upper limb.

Donor nerve	Recipient nerve	Author
XI	SSC (spinati)	Malessy 2004, Bahm 2005
XI	MC (biceps)	Narakas
IC	MC (biceps)	Malessy 1989
IC	AX (deltoid)	Malungpaishrope 2007
U	MC (biceps or brachialis)	Oberlin 2002, Liverneaux 2006
U	R (triceps)	Gilbert 2011
M	MC (biceps or brachialis)	Oberlin 2008
M	R (PIN)	Mackinnon 2007, Bertelli 2010
R	AX (deltoid)	Leechavengvongs 2003, Colbert and McKinnon 2006, Bertelli 2007
MC (brachialis)	M (finger flexion)	Palazzi 2006
ThD	ThL (serratus anterior)	Uerpairojkit 2009
Ph	MC	Gu 1989, Siqueira 2009
Ph	SSC	Sinis 2009
Contralateral C7	Multiple	Gu 1987

Abbreviations: *XI* spinal accessory nerve; *SSC* suprascapular nerve; *MC* musculocutaneous nerve; *IC* intercostal nerves; *AX* axillary nerve; *U* ulnar nerve; *R* radial nerve; *M* median nerve; *ThD* thoracodorsal nerve; *ThL* long thoracic nerve; *Ph* phrenic nerve

nerve over longer distances (more developed in the distal limb part, where the fascicular groups tend to separate to address their end organs).

If a direct coaptation is impossible, then autologous grafts are the best choice, followed by bioartificial solutions like tubes or conduits. Generally excepted in severe war injuries affecting, e.g., the ischiatic nerve with long defects, the dispensable material in the human body (two sural and saphenous nerves, two superficial radial nerves) allows reconstruction of most nerve gaps.

Nerve transfer surgery has been a revolution for the reconstruction of nerve root avulsions or very long gaps. Table 5.1 gives an overview of routine motor nerve transfers performed today.

Nerve transfers are most efficient if they are highly selective and motor/sensory specific and if the muscle target is reached within 18 months (see § on the motor end plate). Their indications actually continue to extend for patients affected by more central nerve pathology (para- and tetraplegia, stroke, spasticity) and within other surgical specialties (urology: bladder innervation; ophthalmology: corneal reinnervation …). This particular field has to be followed as there may be more surprises coming out over the next years.

References

1. Gordon T. Peripheral nerve regeneration and muscle reinnervation. Int J Mol Sci. 2020;21(22):8652.
2. Menorca RM, Fussell TS, Elfar JC. Nerve physiology: mechanisms of injury and recovery. Hand Clin. 2013;3:317–30.
3. Willand MP, Nguyen MA, Borschel GH, Gordon T. Electrical stimulation to promote peripheral nerve regeneration. Neurorehabil Neural Repair. 2016;5:490–6.
4. Zhang CG, Gu YD. Contralateral C7 nerve transfer—our experiences over past 25 years. J Brachial Plex Peripher Nerve Inj. 2011;6(1):10.
5. Gupta R, Chan JP, Uong J, Palispis WA, Wright DJ, Shah SB, Ward SR, Lee TQ, Steward O. Human motor endplate remodelling after traumatic nerve injury. J Neurosurg. 2020:1–8.
6. Brunelli GA. Direct muscular neurotization. J ASSH. 2005;5(4):193–200.
7. Millesi H. Neurolysis. In: Boome RS, editor. The brachial plexus. The hand and upper extremity, vol. 14. New York: Churchill Livingstone; 1997.
8. Assmus H, Antoniadis G, Bischoff C, Haussmann P, Martini AK, Mascharka Z, Scheglmann K, Schwerdtfeger K, Selbmann HK, Towfigh H, Vogt T, Wessels KD, Wüstner-Hofmann M. Diagnosis and therapy of carpal tunnel syndrome—guideline of the German Societies of Handsurgery, Neurosurgery, Neurology, Orthopaedics, Clinical Neurophysiology and Functional Imaging, Plastic, Reconstructive and Aesthetic Surgery, and Surgery for Traumatology. Handchir Mikrochir Plast Chir. 2007;39(4):276–88.

6 Working on Targets: Reconstructive Surgery of Muscles, Tendons, Joints, Bones

J. Bahm, *Surgical Rationales in Functional Reconstructive Surgery of the Upper Extremity*,
https://doi.org/10.1007/978-3-031-32005-7_6

6.1 Muscle Transfer Surgery

6.1.1 Surgical Anatomy and Physiology of a Striated Muscle

Any muscle in the upper limb is a functional unit, assuming a specific contractile motor function. The muscle itself is made of sarcomeres, defined by their intrinsic contractile capacity, responsiveness, and strength, determined by the quality of nerve input, the intrinsic structure and function, and the demanded task.

Sarcomeres arranged in **series** will define the muscle excursion (the contractile or retractile capacity, defined by the sum of individual sarcomere shrinkage), and the sarcomeres arranged in **parallel** determine its strength (the cross-sectional area of a muscle gives an idea about how may sarcomeres work in parallel) (Fig. 6.1). This arrangement cannot be changed by surgical means and is characteristic for each healthy muscle.

In pennate muscle, the muscle fiber/sarcomere orientation is oblique to the line of pull, and this angle will influence the strength transmitted to the tendon according to a simple mathematic law (Fig. 6.2). Muscle might be uni- or bipennate, and this arrangement might be known by microanatomical study, but not by external visualization.

For a given muscle, imbedded in its compartment, surrounded by a fascial limit, the developed excursion and strength are determined by the neural input (type and intensity of motor nerve input) and the task-related output (external demand). Both may vary.

The nerve input will change if the nerve is compressed or traumatized, or if the nerve is altered by surgery (suture, nerve transfer, where a new donor is brought in and will prime the further motor function). The task might change due to variable gravity-related force (standing or sitting position, updrive in the water, existence on moon) or variable task.

Any muscle taken out of its compartment (and thus losing the fascial constraints), either after simple fasciotomy or within the dissection relative to a muscle transfer surgery, will change its properties and especially augment its excursion (see Peckham et al. in Brand [1]).

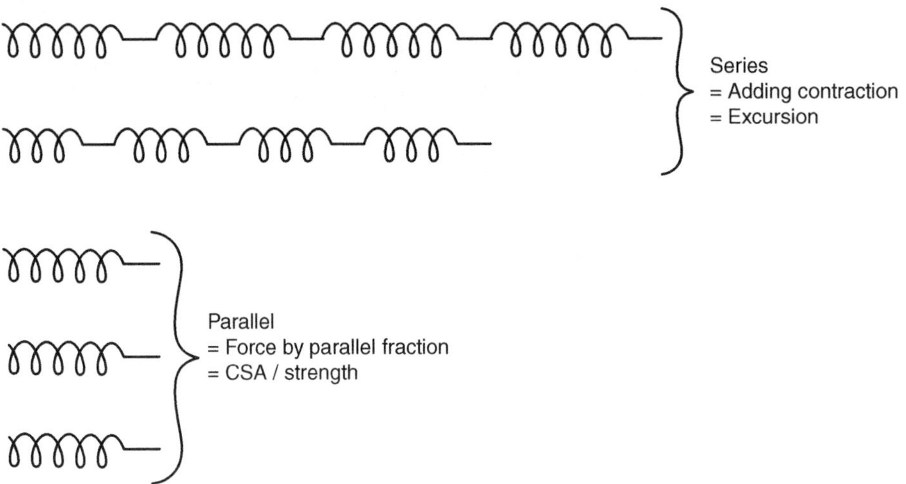

Fig. 6.1 Sarcomeres in parallel and in series

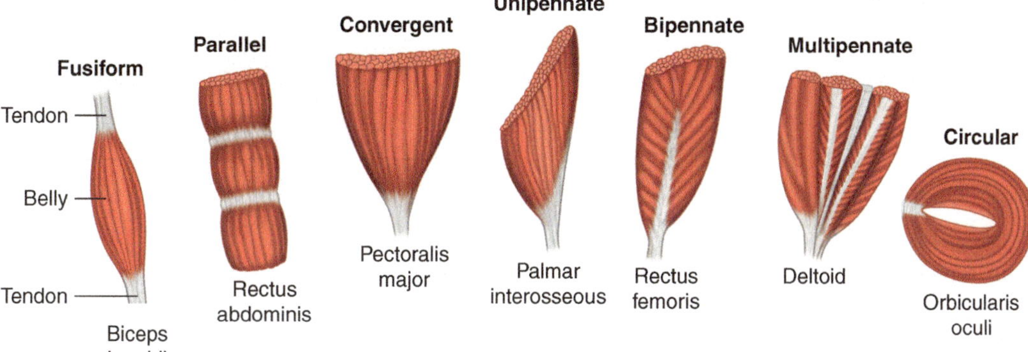

Fig. 6.2 Muscle pennation

6.1.2 Types of Muscle Transfers

More frequent are the so-called *pedicled* flaps, where the neurovascular bundle is respected (eventually just dissected free to enhance an arc of tissue rotation), thus maintaining the vital blood supply to the muscle and the motor nerve control.

A functional *free* muscle transfer implies the dissection and transection of the neurovascular bundle, which will be reconnected by means of microsurgical coaptation to re-establish an arterial and venous blood flow altogether with a motor nerve donor coaptation, which will impose its motor control on the imported muscle. Although the blood supply is restored within the surgery (within a critical time of 6–8 h of intraoperative ischemia time), the nerve function will reappear only after the distal nerve end has regenerated through the proximally coapted donor nerve stump (see Sect. 5.1. on nerve regeneration). Meanwhile, the transferred denervated muscle should and could be prevented from denervation atrophy by regular external electrical stimulation (exponential waves), at least actually applied in the adult patient.

6.1.3 Dissection Technique

Muscle tissue is well vascularized, smooth, but fragile.

It must be handled very delicately, with no rough grasp and no excessive stretching. Prevent intraoperative drying by application of moist gauze or hiding the dissected muscle in a subcutaneous pocket before definitive transposition and wound closure.

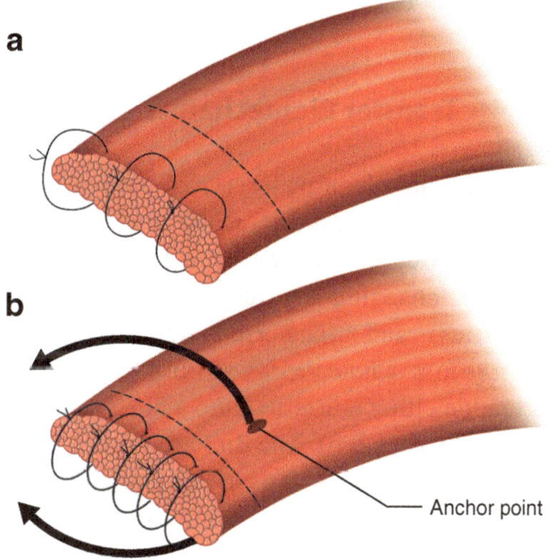

Fig. 6.3 (**a** and **b**) Muscle anchoring suture technique

Any suture of material into a muscle bears the risk of a rupture. Therefore, a line of stay sutures is mandatory (Fig. 6.3a) before anchoring further sutures, e.g., for the tendon or bone fixation, **behind the line of stay sutures** (Fig. 6.3b).

Any transverse cut through a muscle, even when repaired (sutured), will induce local fibrosis and definitely affect the contractile property of this muscle.

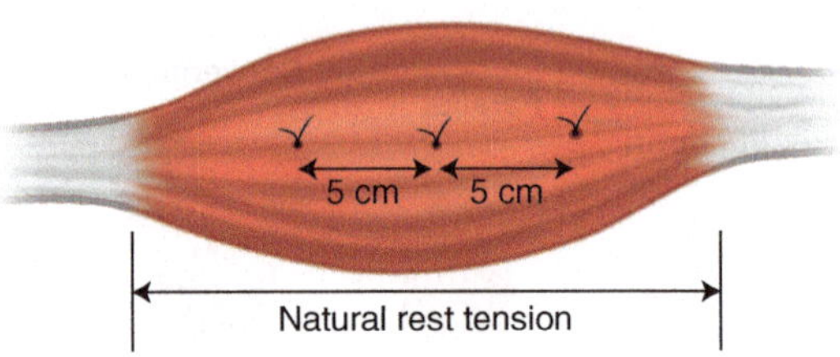

Fig. 6.4 Tension stay sutures

In any muscle transfer surgery, appropriate tension of the muscle is one of the major factors of postoperative functional success. The best tension may be judged on spaced stay sutures, reflecting the preoperative tensile status, e.g., applied every 5 or 10 cm (Fig. 6.4) and/or by manual appreciation of the elastic muscle behavior (excursion) related to the target needs.

6.1.4 Transfer Biomechanics

Any real muscle transfer ("real" when compared to a tendon transfer) means that the donor muscle is dissected out of its original bed (thereby changing its anatomical excursion properties) and transferred under the skin (mostly in a subcutaneous gliding tunnel, with its tendon either reinserted into a bone or woven into another tendon). The direction given to the muscle, the type of soft tissue bed, and the chosen length will affect the target function (Table 6.1). Additionally, the target position relative to the nearby joint and the type of tendon fixation (inducing more or less postoperative adhesions) will also affect the functional outcome.

Table 6.1 Factors influencing target function [2]

- Available muscle excursion (dissected free of the fascia or not)
- Length-tendon characteristics of the individual muscle
- Total force available: active component (contraction) and passive internal properties (excursion)
- Active component: contraction of the muscle-innervation-neural recruitment
- Passive component: internal muscle properties and architecture—pennation, fiber types, and proportion
- Others: direction of pull, tendon gliding tissue, antagonistic activity, and regulation [2]

Considering the possible lever arm over the target joint is one of the most important biomechanical factors to be considered in the upper limb, especially at the shoulder level (Fig. 6.5).

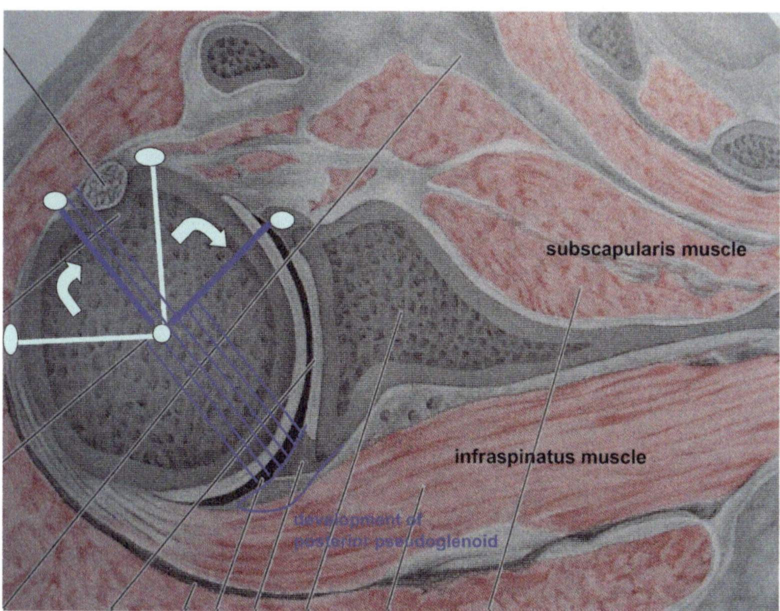

Fig. 6.5 Force vectors around the humeral head

6.2 Tendon Transfer Surgery

The term "tendon transfer" means the transposition of a functional tendon-muscle unit into another tendon (rarely into a bone insertion) without dissecting out the whole muscle body, thus leaving its local properties unchanged. It is popular within the hand and wrist, e.g., when replacing the EPL tendon by an EIP transfer after a rheumatoid or traumatic rupture of the EPL tendon.

Here, the main concerns are a proper tendon suture and the prevention of adhesions (need for early mobilization).

Tendon sutures are here commonly of Pulvertaft type, needing a special pull-through clamp (Fig. 6.6). The technique uses either several pull-throughs in perpendicularly changing directions (to avoid the creation of a fusion hole) or side-to side fixation after one pull-through ([3], experienced in tetraplegia surgery, Fig. 6.7).

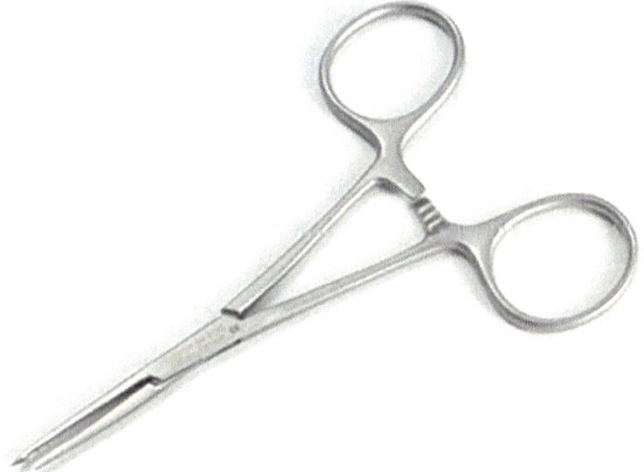

Fig. 6.6 Specific instrument for Pulvertaft-type tendon weaving and fixation

The creation of a proper gliding bed is frequently neglected, mostly due to local tissue lack, and early mobilization is not always feasible due to patient characteristics (low age, low compliance).

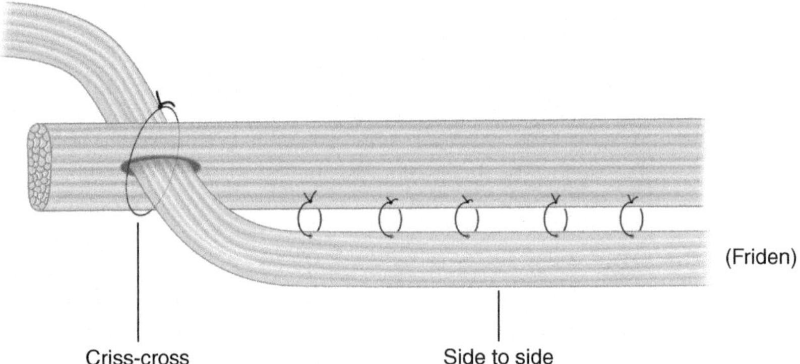

Fig. 6.7 Friden's side-to side technique

6.2.1 Tendon-Bone Fixation

Due to numerous industrial developments, we moved away from simple periosteal sutures or transosseous sutures to the new world of *bone anchors*, mainly developed in the field of orthopedic (shoulder and wrist) surgery. Today, a cortical bone anchor armed with nonabsorbable, reinforced filaments will allow a good-quality suture with due tension regulation. Several systems exist to bring the tendon end into the bone to allow a tighter tendon-bone interaction, reproducing a natural enthesis (for example to refix a ruptured biceps tendon into the proximal radius).

6.3 Bone and Joint Corrections

The major involved joints are the glenohumeral joint (humeral head rotation), the elbow (olecranon hinge and radial head translation-rotation), and the complex wrist joint.

The diaphyseal bones addressed in reconstructive surgery are the humerus and radius, for (de)rotational osteotomies and osteosynthesis either by open reduction and internal fixation (ORIF, mostly by locking plates) or by external fixation.

The most interesting parts are the humeral head (and its influence on the glenoid shape and version) and the behavior of the radial head.

All joint partners in children are at risk of subluxation, dislocation, and morphologic changes due to transmitted strength or torque changes.

Muscle disbalance is supposed to be the primary steerer, but physiologic motion under altered constraints in a progressively incongruent situation seems to play another important role.

Morphologic changes only happen in the growing joint and are detailed in Chap. 7.

In the adult patient, the already established muscular imbalance has pushed the joint partners into a new equilibrium, which might be unstable, fixed, dysplastic, and sometimes painful.

In the growing child, there is a specific dynamic described by Pauwels, explaining how a change in forces and torques will influence histogenesis and alter joint surfaces and their shape, including the joint base (like the glenoid neck). Here, our early action may be preventive (concerning deformity) and a re-established joint congruence may insure a better development, function, and lower risk of late arthritic changes.

But obviously, we are unable to screen all children early enough; we cannot identify all pathogenic factors and correct them sufficiently, and we still see late deformities although an early and complete treatment occurred. Our strategy still is insufficient.

6.3.1 Contracture and Posture

Any muscular imbalance, either primarily due to the unequal nerve damage affecting the antagonists or secondarily residual after a treatment, produces a soft tissue contracture mainly of the joint capsule and adjacent soft tissue by shrinkage, fibrosis, and reduction of the passive range of motion, which turns with time into a fixed posture.

The contracture is addressed both by direct methods of passive mobilization (physiotherapy and orthoses) or surgical release and by a trial of restauration of the incriminated imbalance.

The posture is often discovered lately, e.g., at school age, and the joint status may not be reversed. If an acceptable joint motion exists, the correction is shifted to the adjacent diaphysis and for example a rotational osteotomy of the humerus is proposed.

For the subluxated radial head, a corrective 3D osteotomy with a wedge-shaped resection may allow to restore some humero-radial congruence, although the capacity of prosupination may still be very limited.

6.3.2 Joint Dysplasia

The malformed cartilage cannot be corrected by direct surgery. Only the underlying bone plateau (glenoid or radial head neck) may be addressed to change their position and length, in order to reduce the difference to normal anatomy and biomechanical behavior.

We address this complex and dynamic deformity in the following chapter.

6.3.3 Corrective Diaphyseal Osteotomies

The rotational limb segment position may be changed by performing a transverse osteotomy of the underlying long bone diaphysis, rotational arrangement of the distal segment, and ORIF by plate or external fixation. The following procedures are commonly performed in adults and older children when more dynamic procedures, like muscle-tendon transfers, are no longer indicated or excluded due to donor weakness:

1. **Humerus**: Rotational osteotomy in lateral rotation (generally 30–50°) to improve the ROM into lateral rotation of the shoulder; rarely an osteotomy and medial rotation are performed in those situations where an anterior shoulder release repositioned the humeral head into the anterior (true) glenoid, thereby limiting the passive ROM in medial rotation of this humeral head. The medial rotation range is thus recovered by a medial rotation osteotomy of the distal humeral diaphysis, mostly simultaneously within the release procedure.
2. **Radius**: Pronating osteotomy to correct a static supination contracture of the forearm and to convert this ugly, nonfunctional position of a beggar hand into a more natural, and acceptable pronation position (bringing the hand palm onto the working table).
3. Ulnar lengthening versus radial shortening in radial head problems (see tridimensional surgical correction of radial head subluxation).

6.4 Contracture Release

6.4.1 What Are Soft Tissue Contractures?

The term "contracture" should not be used in any consideration about the contraction status or contractile capacity of living and innervated muscle tissue.

The contracture means stiffening of soft tissue (muscle, ligament, joint capsule) by means of collagen fiber invasion and organization, resulting in a decrease of a passive range of motion.

It is through this decrease of passive mobility that we suspect a contracture, which then might be confirmed, e.g., by ultrasound imagery. Any long-lasting contracture in the growing child may affect the growth process of various tissues, and any contracture will not resolve by itself, but worsen with time.

A soft tissue contracture affecting tissues around a joint will impair the joint's mobility and interfere not only with the active motion, but also

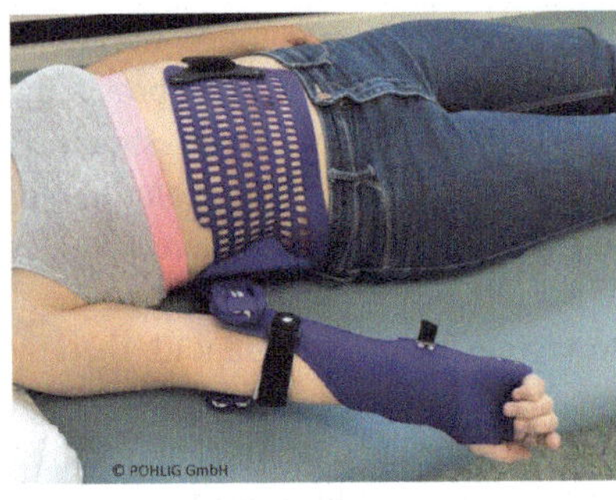

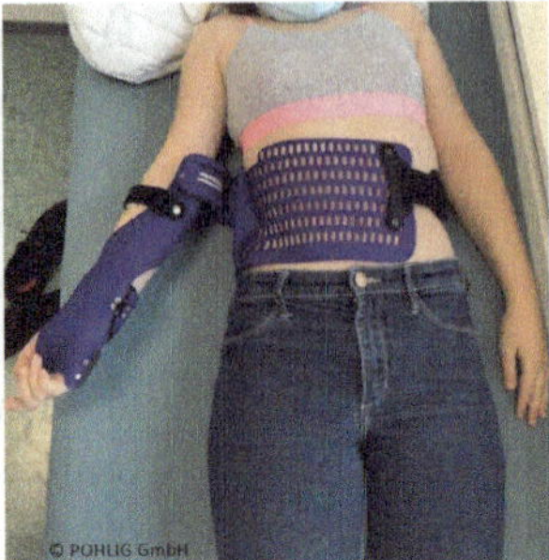

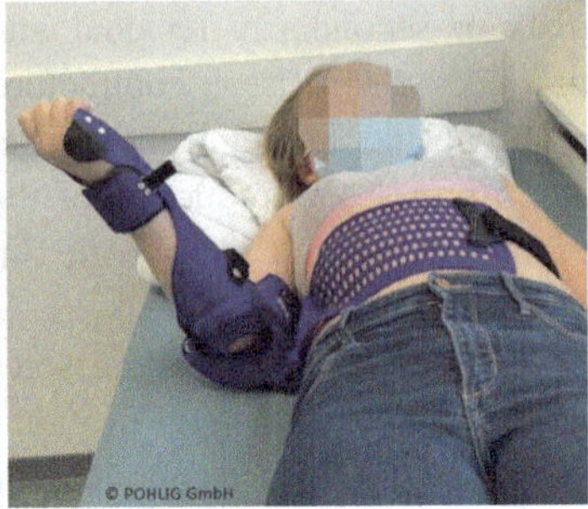

Fig. 6.8 Pohlig-Bahm orthosis for the OBPP upper limb to correct the medial rotation contracture of the shoulder and supination contracture of the forearm. © POHLIG GmbH

with positioning at rest, thereby allowing the development of an abnormal posture, becoming more fixed over time.

The contracture may be a morphologic reality or a pathophysiologic concept (e.g., in physiotherapy or osteopathy), and ideally both should concur and be addressed by therapy in a way that the impaired active function may return.

How does connective tissue (collagen fibers) alter muscles, gliding tissue, ligaments, and joint capsules? Post-traumatic or postoperative healing is associated with the development of adherences, shortening of so-called *connective* tissue due to immobilization, scar formation … even muscles immobilized in elongated or shortened positions will vary their sarcomere length, and thus muscle "shortening" by reduced sarcomere length may contribute to the surrounding shrinkage process.

There is so far no drug interfering with fibrosis, adherence, and soft tissue shrinkage. The medical attitude therefore works through prevention (mobilization, corrective postures given by adapted orthoses), passive stretch exercises, and surgery.

6.4.2 Prevention by Orthosis

Obstetric or cerebral palsy in the infantile upper limb is frequently associated with similar contracture patterns, addressed by regular physiotherapy and adapted orthoses.

The shoulder is frequently adducted and medially rotated, the elbow flexed (thus risking an extension deficit), the forearm pronated, and the wrist flexed. We therefore participated in the development of a modern, light, versatile, and adaptable soft material upper limb orthosis manufactured individually and addressing all joints of the upper limb specifically, either in a preventive or in a curative intention (Fig. 6.8: 4D orthosis).

This orthosis allows a posture correction at night and fits all modern requirements.

6.4.3 Surgery

(See also the topographic order in Sect. 7.3.)

Surgery of a contracture is coined by the term "release." It consists of excision or division of the fibrotic tissue, eventual correction of secondary tissue changes (local bone overgrowth, joint incongruence), or dynamic imbalances, which might be responsible for the local posturing and development of contractures (shoulder: medial vs. lateral rotators; elbow: biceps vs. triceps activity).

Of course, the release surgery also leaves adhesions and scars, and the postoperative course must help to maintain the intraoperatively gained passive ROM and prevent recurrence of the same contracture pattern. Therefore, physiotherapy must specifically address the individual issues after a contracture release, i.e., the risk factors for recurrence, the idiopathic fibrosis, and the regain of an enhanced activity and active ROM, which ideally should "protect" the augmented passive ROM.

When there is spasticity, the muscle slide or release techniques might be confounded with the abovementioned description.

Surgery of the spastic muscle concerns a still contractile organ, which the surgery should not delete. Here, the release concerns the detachment of a proximal muscle-bone attachment (e.g., the flexor-pronator mass), a tendon lengthening, or a selective motor hyponeurotization—all of these are aimed to reduce the muscle tone (there is no such "contracture") while preserving the functional aspect of muscle contraction. This surgery allows to participate in contracture release, but also induces reduction of muscle strength.

Any muscle afflicted by a spastic deregulation is invaded by soft tissue over time (loss of elasticity, increased impedance), associated sometimes with denervation atrophy and fatty infiltration. Assuming those complex changes in muscular morphology, there is so far no reasonable intramuscular surgical procedure to improve the contractile quality (a contrario: all muscle incisions and sutures producing new scarring are to be limited or avoided).

References

1. Brand PW, Hollister A. Clinical mechanics of the hand. 2nd ed. St Louis: Mosby Year Book; 1993.
2. Peckham PH, Freehafer AA, Keith MW. The influence of muscle properties in tendon transfer. In: Brand PW, editor. Hollister a: clinical mechanics of the hand. 2nd ed. St Louis: Mosby; 1993.
3. Fridén J, Tirrell TF, Bhola S, Lieber RL. The mechanical strength of side-to-side tendon repair with mismatched tendon size and shape. J Hand Surg Eur. 2015;40(3):239–45.

7 Growing Joints: Disbalance, Incongruence, Dysplasia

J. Bahm, *Surgical Rationales in Functional Reconstructive Surgery of the Upper Extremity*,
https://doi.org/10.1007/978-3-031-32005-7_7

7.1 Physiology of Joint Development and Pathophysiology in Muscular Disbalance

7.1.1 Functional Histogenesis Described by Pauwels

Considerations about shoulder biomechanics based on Pauwels' work extended from the hip to the shoulder are described in detail in specific books. A particular point is that bone and joint surface deformation (here for the glenoid) can be explained with Pauwels' hypothesis on "causative" **functional histogenesis** considering a dysplastic adaptation of bone structures under chronic dynamic pressure constraints.

In his studies on hip joint biomechanics, Pauwels established vector constructions for the forces applying onto the femoral head in both frontal and transverse sections, to study the weight-bearing and structural anatomic adaptation within the femoral head and neck. He included muscle force amplitudes and gait dynamics.

We tried to apply his construction in the transverse plane to our considerations on the force balance between medial and lateral rotators at the glenohumeral joint. Although vector origins and directions could be applied comparatively without problem, the main limitation consisted of the lack of data concerning shoulder motion dynamics in different directions (e.g., medially or laterally oriented combined flexion-adduction or flexion-abduction movements) and the muscle strength of normal or weakened medial and lateral shoulder rotators in children. Therefore, it was impossible to add force vectors of unknown size, although directions and origins were known. At this moment, no further conclusions may be made from this type of force vector constructions applied to the glenohumeral joint in children.

Kummer, who had worked with Pauwels, added two interesting concepts about the pressure inside the glenohumeral joint:

1. The combination of gravity (weight of the arm) and muscle power of the deltoid results in a vector in line with the joint, concentrating the force between the two joint components (the humeral head and the glenoid). He does not mention if this remains true with a weakened deltoid (palsy) or a muscular imbalance. We suggest that the clinical observation of a posterior glenoid deformation corresponds to a different force axis.
2. The relationship between the humeral head and the glenoid with a greater radius for the first as a "balanced incongruence" where the contact pressure is first applied to the glenoid rim and does not touch the deep center of the glenoid. He cites Pauwels' work who analyzed tension lines of the glenoid concavity to confirm this concept. If the forces are oriented more posteriorly, the posterior rim will be deformed and the concavity will change into a posterior pseudoglenoid or a neoglenoid, as described in the MRI series of affected OBPP children.

7.1.2 The Humeral Head

Imagine a ball, suspended in space. Downwards, this sphere is prolonged through the neck into the humeral diaphysis, pointing in the direction of gravity. But otherwise, every movement is possible, due to a lax envelope and tendon attachments (Fig. 7.1) allowing rotation (inwards and outwards: medial and lateral), abduction and adduction, and forward and backward flexion (ante- and retropulsion).

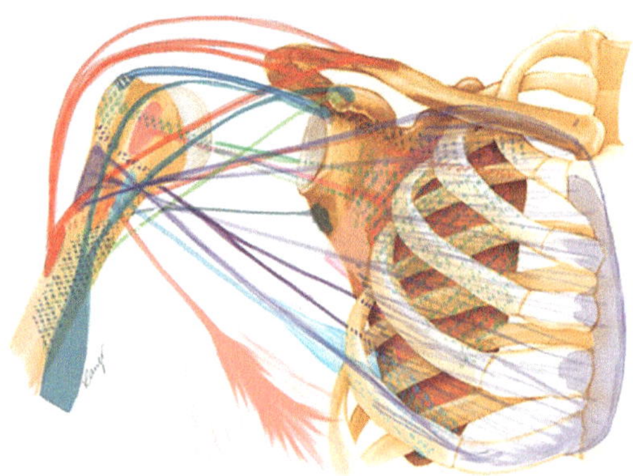

Fig. 7.1 The humeral head and its "suspension" (Langer)

The socket, called glenoid, offers little constraints, and the passive and active range of motions in the growing child and young adult are just incredible.

That is our glenohumeral joint, a very mobile unit, suspending the whole upper limb, held by ligaments and a lax capsule. It is the master of freedom, but also of tightness (in frozen shoulder), guaranty for stability in all motions, but subjected to dysplasia, when the muscle balance in the growing period is altered.

7.1.2.1 Large Cartilage: Late Ossification

Recent MRI data show [1] that in the humeral head, preossification centers are already seen at 2 and 4 months of age, where secondary ossification centers appear in the medial humeral head and greater tuberosity at 4 and 10 months, respectively. Fusion begins by 3 years and completes by 13 years only!

How long are the humeral head and the glenoid really "plastic" and can correct an incongruence?

In the resting upright position one holds over the day, abduction/adduction and ante- and retropulsion are at neutral—only the rotational status may vary due to imbalanced action of the medial or lateral rotators. Their tendons and tendomuscular junctions come tight to the capsule and influence directly the head positioning and its influence on the glenoid.

In the adult, these partners are resistant; but in the young child, the smooth cartilage may get deformed, and the neck angulation both of the glenoid AND of the humeral head may adapt and change.

A frequent issue happens in upper obstetric palsy, when the lateral rotators (overall the infraspinatus muscle) are weakened and recover more slowly and often incompletely, while the medial rotator (the subscapularis muscle) remains often strong and not counteracted: a medial rotation posture occurs, where the humeral head is translated dorsally, and the posterior glenoid is more in charge (the anterior part discharged), allowing adaptive growth and glenoid dysplasia, increased retroversion of the glenoid neck, and adaptive (less) retroversion of the humeral head, with at endpoint a dorsally dislocated humeral head facing a distorted pseudoglenoid and an empty glenoid cavity undergoing atrophy (Fig. 7.2).

Glenoid deformity has been classified according to newer image diagnosis (CT, MRI) mainly considering the shape of the glenoid fossa: concave (normal), flat, biconcave (with a posterior pseudoglenoid), and even convex [2, 3].

type I normal glenoid - less than a 5-degree difference in retroversion compared with that on the normal, contralateral side

type II minimum deformity – more than a 5-degree difference in retroversion compared with that on the normal side, with **no posterior subluxation** of the humeral head

type III moderate deformity - **posterior subluxation of the humeral head**, defined as less than 35 per cent of the head anterior to the bisecting line

type IV severe deformity including a **false glenoid**

type V severe flattening of the humeral head and glenoid, with progressive or **complete posterior dislocation of the head**

type VI dislocation of the glenohumeral joint

reference:

Waters PM, Smith GR, Jaramillo D. Glenohumeral Deformity Secondary to Brachial Plexus Birth Palsy. The Journal of Bone & Joint Surgery: 1998 80 (5) 668-77.

Fig. 7.2 Glenoid deformation

The retroversion of the glenoid and of the humeral head may be calculated if the image includes the elbow joint as a reference line.

The more or less spherical shape of the humeral head and the shape of the acromion may also be investigated.

We have been able to show that the use of a medially rotated arm in daily activities will produce forces at the glenohumeral joint level, which will push the humeral head backwards and thereby maintain or worsen the position of posterior subluxation and progressive glenoid dysplasia [4].

The surgical options to rebalance the rotators, to relocate the joint (re-establishment of joint congruence), and to correct the version angles are addressed in Sect. 7.2.

There is so far no solution to interfere with the cartilage dysplasia itself, and even in the adults suffering from this condition, arthroplasty is not an option, due to frequently insufficient motor control.

7.1.3 The Radial Head

The radial head participates "twice" in the complex elbow joint: in the constrained hinge joint of flexion-extension, it allows the radius to approach or leave the humerus; in prosupination, through the proximal radioulnar joint, a rotation AND translation of the radial head make the radius turn around the ulna, around an oblique axis from the ulnar styloid to the radial head (Fig. 7.3).

Flexion–extension is a power movement involving on the extensor side the triceps, on the flexor side the brachialis muscle as a starter (when the angle is zero), and the biceps muscle as the powerhouse. The brachioradialis muscle seems to play a modulating role.

Prosupination is a complex movement occurring through the proximal and distal radioulnar joint, described by some authors as a large bicondylar joint made by those two joints united by the interosseous membrane.

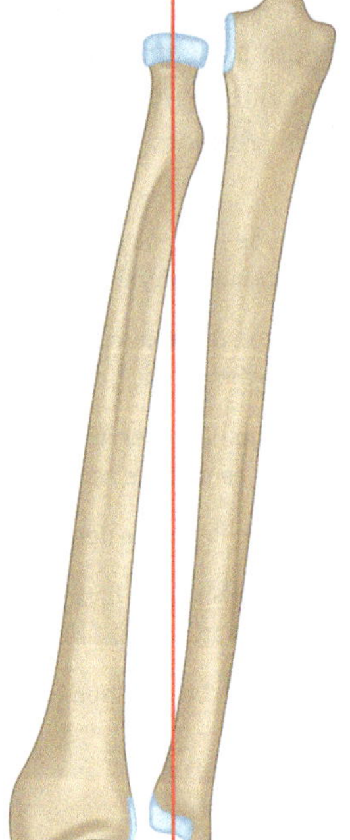

Fig. 7.3 Radial head movements

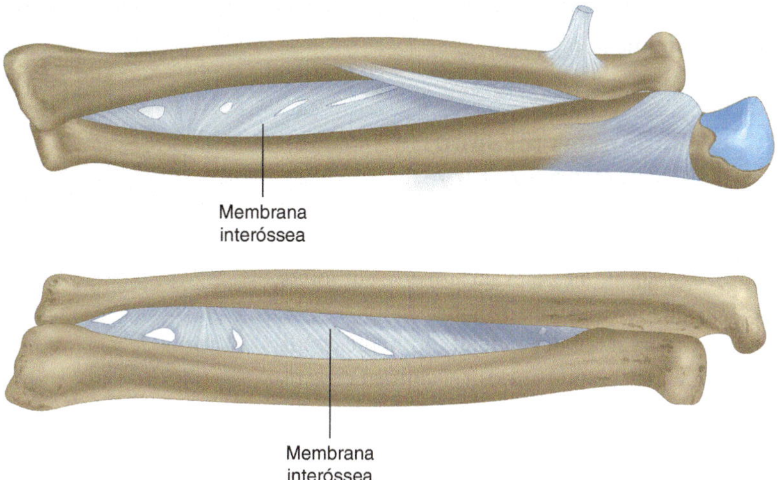

Fig. 7.4 Interosseous membrane—the concept of the bicondylar joint

The latter is an important stabilizer with a subtle anatomy (Fig. 7.4), and its influence in pathologic conditions is not always clearly understood (Essex-Lopresti injuries).

In theory, a disbalance may occur in the flexion–extension plane if one of the partners is weakened (development of an elbow flessum or extension lag) or if the brachioradialis is affected.

The stability or congruence of the radial head occurs in post-traumatic or palsy situations, and both anterior and posterior subluxations are described. Any major incongruence of the radio-humeral joint will reduce the local constraint on the proximal radius and allow excessive and distorted growth of the proximal bone, including dysplasia of the radial head cartilage and proximal bowing. In late cases, not only the prosupination may be affected, but also the active and passive elbow flexion may be reduced.

The exact pathophysiologic agent is not always identified, and thus the treatment is seldom etiologic.

In obstetric palsy, the radial head may be found in anterior subluxation, and one incriminates the "excessive" pull of the distal biceps tendon inserted on the radial tuberosity.

We observed that in most of the cases presenting with this anterior type of subluxation, there is an associated supination contracture—meaning that supination and forward movement (translation) of the radial head go together, which is found in physiologic conditions: when supinating, the radial head slightly moves forward, and in pronation, it returns into the joint cavity.

One might suspect that the altered range of prosupination in these cases underlies an excessive and permanent tendency to anterior translation, which the annular ligament cannot restrict with time. Also, we should add that elbow flexion activity normally occurs with a supinated forearm and that the biceps muscle itself is supinating, meaning that the most frequent elbow movement—active flexion—bears an inherent risk of *anterior* radial head translation.

Corrective surgery at its best should rebalance early but is frequently reduced to relocate the radial head once the subluxation is diagnosed and to diminish the biceps tendon action onto the proximal radius, by transferring the distal biceps tendon unto the proximal ulna (see Sect. 7.3.2.2 and 7.3.2.3).

The ossification of the partners of the radiocapitellar joint happens between 1 and 3 years—

remodeling thus requires an early re-establishment of joint congruence.

7.2 Rationales for Joint Correction Surgery

After having set the anatomic and biomechanical knowledge about the relevant upper limb joints, we may actually establish some rationales for corrective surgery, independently from the conditions which are responsible for disturbed congruence and muscular imbalance.

In the growing child, everything should be done to allow a normal growth, which means for the joints balanced forces of the antagonistic muscles, congruence of the anatomic partners, and an unrestricted soft tissue environment.

In the population we consider, even in adult patients, joint replacement is not a solution, because of sometimes severe dysplasia and also unbalanced or weak muscles. Direct surgery on the cartilage is not possible, and all what surgery can offer today is to correct growth disturbances or bone irregularities counteracting congruent joint development.

Corrective joint surgery, although aiming to restore the form to improve the function, always bears the risk to improve the X-ray or CT scan appearance of a joint at the price of reducing the passive ROM by scarring and adherences, not always further relieved by physiotherapy.

At the **glenohumeral joint** level, corrections may be done with a focus on the glenoid neck version (open wedge osteotomy to reduce the posterior tilt) or its global shape (addressing the pseudoglenoid or posterior cavity). The humeral head version may also be addressed, and the diaphyseal rotation corrected in both directions by a (de)rotational humeral shaft osteotomy.

In small children, a lot of prevention may be done by passive exercises and night splints, rebalancing the humeral head when the arm is adducted and turned into lateral rotation (Fig. 7.5 cf. Sect. 6.4.1).

Considering the **radial head**, resection is not an option in the growing child, as there is a high risk to develop compensatory wrist changes with severe ulna plus variance, probably due to adjust-

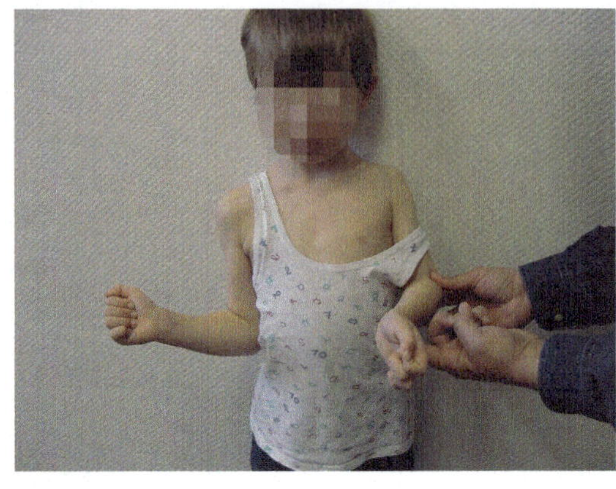

Fig. 7.5 Passive exercise of lateral rotation with the arm in adduction

ments through the still supple interosseous membrane.

The subluxated or dislocated radial head loses the limiting and stabilizing contact to the capitellum, responsible for proximal shaft overgrowth, reduction of the physiological shaft curvature, and progressive atrophy and fibrous dysplasia of the covering cartilage. It is thus important to detect incongruence early, and eventually to try restoration of congruence by means of open radial head reposition, requiring frequently a shortening proximal shaft osteotomy (correcting the overgrowth) associated with a three-dimensional correction to allow the radial head to align naturally with the capitellum. During surgery, one must associate the knowledge of the influence of prosupination with the translation of the radial head: pronation brings the anteriorly subluxated radial head back posteriorly into the fossa of the capitellum (where the opposite should be true in a posterior displacement).

Interestingly, in our cases suffering from obstetric palsy presenting with anterior radial head subluxation, there is frequently also a severe supination contracture of the forearm. We believe that it is not the pull of the distal biceps tendon onto the radial head which allows the unrestricted anterior displacement, but rather the uncompensated severe supination deformity and contracture, exerting anterior translation onto the radial head and weakening the annular ligament by distension and proximalization. In some operations, we were able to reposition and maintain the anteriorly subluxated radial head by a rerouting surgery onto the biceps tendon (Zancolli-Grilli procedure), which after the L-shaped section of the biceps tendon and pronation of the radius allowed full reintegration of the radial head into the fossa.

This type of surgery unfortunately has a very inconsistent long-term outcome, as over the years, the risk of recurrence of the subluxation is high, obviating that not all conditions are stabilized to maintain joint congruence.

One aspect may be insufficient planning of the 3D corrective osteotomy. Preoperative templates like the *Materialise technology* could be applied in the future, with bilateral preoperative CT scan, creation of a 3D model of the corrected proximal radius, and thus clear guides on how to perform the osteotomy to obtain exactly this mirror anatomy, considering the healthy contralateral side (Fig. 7.6).

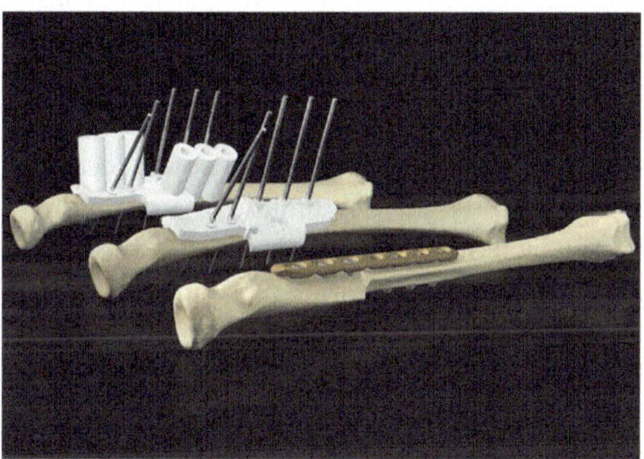

Fig. 7.6 Materialise templates

Another problem is our insufficient knowledge of the local forces, co-contractions between biceps and triceps, regulative influence of the brachioradialis, and role of impaired prosupination—all that in a chronic time perspective.

Also, the radial head position is not accessible to corrections by splints or orthosis.

7.3 Current Procedures in Topographic Order

7.3.1 Shoulder

7.3.1.1 Contracture Release

The release of soft tissue contractures around the glenohumeral joint is aimed to restore the rotational balance and increase the passive range of motion, particularly in lateral rotation.

According to the location of the restriction, we distinguish anterior, inferior, and posterior contractures (Fig. 7.7), which affect the passive ROM in different ways. The anterior restriction clearly limits the capacity of lateral rotation of the adducted arm; the inferior subcutaneous fibrotic band located on the lateral proximal border of the latissimus tendomuscular junction pulls the scapula in outward rotation when the arm is abducted (thereby limiting the active and passive abduction, as the scapula-humeral rhythm is disturbed by the early rotation of the scapula), and the so-called posterior contracture seen at the posterior aspect of the glenohumeral joint just reflects

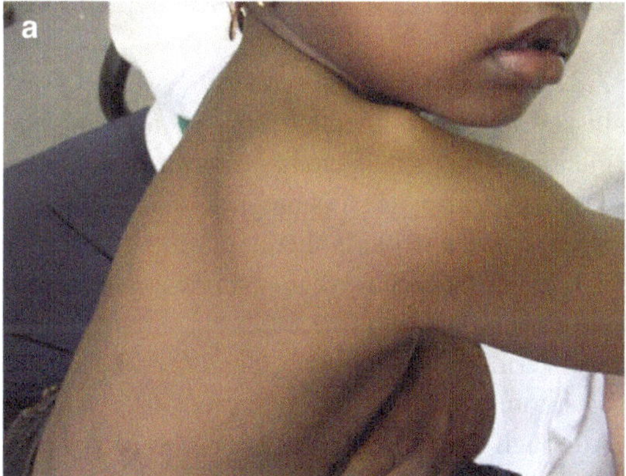

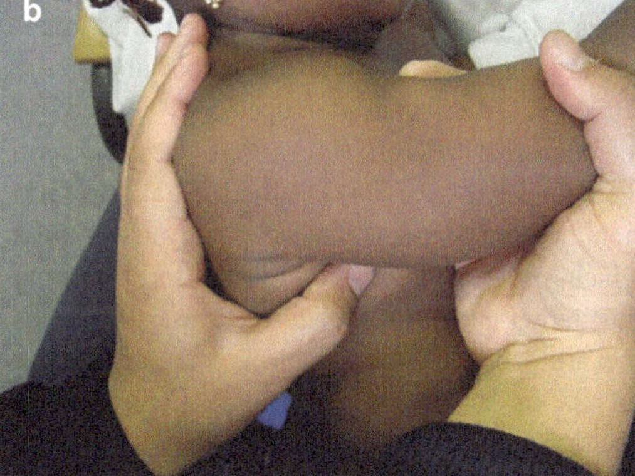

Fig. 7.7 Different types of contractures at the shoulder level. (**a**) visible tilting of the scapula, (**b**) passive stretching exercise

severe glenohumeral dysplasia with concomitant joint incongruence, which is responsible for the humeral head to lever on the posterior glenoid, thereby tightening this posterior soft tissue wall.

So far, I have never observed a severe abduction contracture of the shoulder needing a corrective surgery (muscle release).

7.3.1.2 Anterior Release

This is the most frequent, rewarding but never totally "standard" procedure I have performed in sequelae of obstetric palsy.

When the medial rotation contracture of the shoulder is not relieved at the age of 2 years (pLR ADD >30°), by means of regular physiotherapy, orthosis, botulinum toxin into the subscapularis muscle, or early nerve transfer to neurotize the suprascapular nerve, the release surgery is indicated [2]:

Anterior vertical deltopectoral approach, division of the subcutaneous fat and respect of the cephalic vein which indicates the deltopectoral intermuscular groove, identification of the coracoid process and the conjoined tendon of pectoralis minor and coracobrachialis, section of the coracohumeral and coracoacromial ligament at the lateral border of the coracoid (which sometimes may be sufficient to release), eventual lateral or subtotal resection of a frequently overgrown coracoid (as the humeral head moves posteriorly, the less occupied anterior space may be occupied by the unrestricted growth of the coracoid process), and through this space identification of the subscapular tendon which should be thoroughly dissected (including the resection of the fatty capsular tissue at its superior border, filling the rotator interval) are done. Then, as a last step, the subscapular tendon is either incised (superior V-shaped) or lengthened with a classic step-cut incision, with the lateral flap superior and the medial flap inferior, thus preserving the subscapular muscle mass which spreads more medially-inferiorly. During this stepwise release procedure, the assistant holds the arm adducted and controls the progressive release in lateral rotation and the concomitant anterior movement of the humeral head. Once the joint is free and the glenohumeral congruence restored (the humeral head moves from its posterior position with contact to the posterior pseudoglenoid into the original anterior one, with the hypoplastic true glenoid), the degree of joint dysplasia may also be checked by repetitive movements in lateral and medial rotation. If the humeral head "springs" with a click, dysplasia is considered to be clinically relevant and one must then decide either to leave the head in its posterior location (and perform a lateral rotation osteotomy of the humeral shaft—now or later) or to relocate the head anteriorly, thereby limiting the medial rotation, which sometimes will need a concomitant medial rotation osteotomy of the humeral shaft.

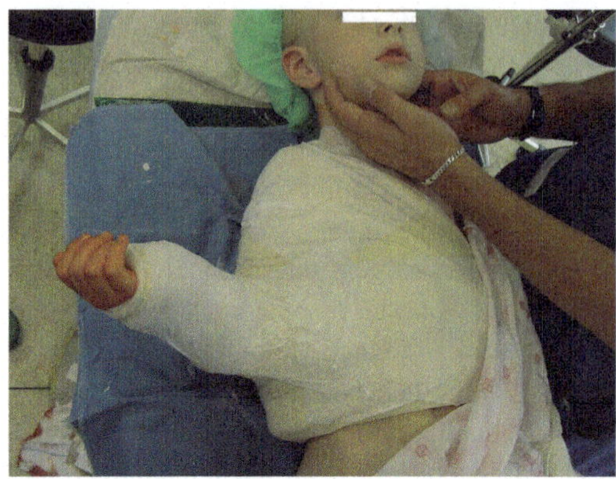

Fig. 7.8 Immobilizing plaster after anterior release surgery

After soft tissue and skin closure, a thoracobrachial plaster in lateral rotation of the arm is maintained for 6 weeks (Fig. 7.8).

7.3.1.3 Inferior Band Resection

It is rarely indicated as the functional benefit is minor and the scar is prone to local hypertrophy.

7.3.1.4 Glenoid Osteotomy

In severe glenohumeral dysplasia, not only the glenoid is shaped differently (convex, flat, biconcave), but also its version, i.e., the inclination relative to the neck, is increased in the posterior direction. In recent years, surgical techniques have been designed to correct this angulation by a posterior glenoid approach and opening wedge osteotomy of the neck to correct this version. Although the pre- and postoperative MRI scans show a good correction, the correlation with improvement of passive range of motion is so far not so beneficial (literature).

7.3.1.5 Muscle Transfers

Muscle transfers aimed to improve the active lateral rotation like the transfer of latissimus dorsi and/or teres major tendon are frequently done as a second procedure, once the passive ROM improves (and if the active motion did not improve accordingly within one postoperative year), but are done in some specific situations, especially in older children, within the same procedure.

It is debated if early muscle transfers really maintain or improve the rotational balance and thereby decrease or inhibit the development of the joint dysplasia [5].

7.3.2 Elbow

7.3.2.1 Anterior (Flexion) Contracture Release

Anterior approach and L-shaped step cut in the tightened and fibrotic brachialis muscle: The biceps tendon is rarely involved, neither the lacertus fibrosus. Anterior capsulotomy rarely adds value, and in older children, one has to accept the growth disturbance with an anteverted humeral condyle complex. Some colleagues correct this deformity by a supracondylar opening osteotomy with a plate osteosynthesis.

7.3.2.2 Surgery of the Radial Head

Relocation of an anteriorly subluxated radial head by means of an open, three-dimensional resection osteotomy and osteosynthesis of the proximal radial shaft: A major issue is the risk of re-dislocation and the quality of residual cartilage of the radial head.

In this surgery, the distal biceps tendon is transferred to the ulna and the annular ligament is reconstructed, eventually using a strip of triceps fascia.

7.3.2.3 Surgery of Prosupination

Rerouting of the supinating distal biceps tendon in a pronating way (Zancolli-Grilli—see Bahm et al. 2002, Figs. 7.9 and 7.10) or of the brachioradialis muscle around the radial diaphysis is done, to enhance either active pronation or supination (Özkan, Fig. 7.11)—provided that the passive ROM in prosupination is near normal (integrity of the proximal and distal RUJ and the interosseous membrane IOM). Otherwise, only a static procedure with osteotomy of the radial shaft in either pronation or supination is feasible (and will just change one fixed posture into another one, more acceptable or functional).

In patients suffering from spasticity, Gschwind [6] established a clear strategy according to the gradual presence of usable active and passive ROM (see Table 7.1). Here, a progressive release of spastic muscles combined or not with tendon transfers is proposed in a staged manner. Also, Özkan's technique [7] applied in spastic pronation contractures, and in an extended variant, may simultaneously correct the spastic ulnarly deviated wrist.

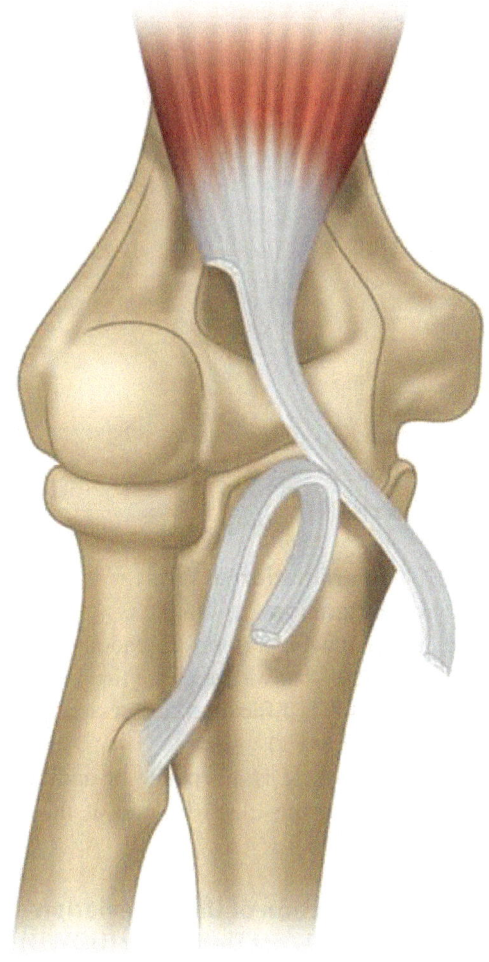

Fig. 7.9 Zancolli-Grilli, step 1

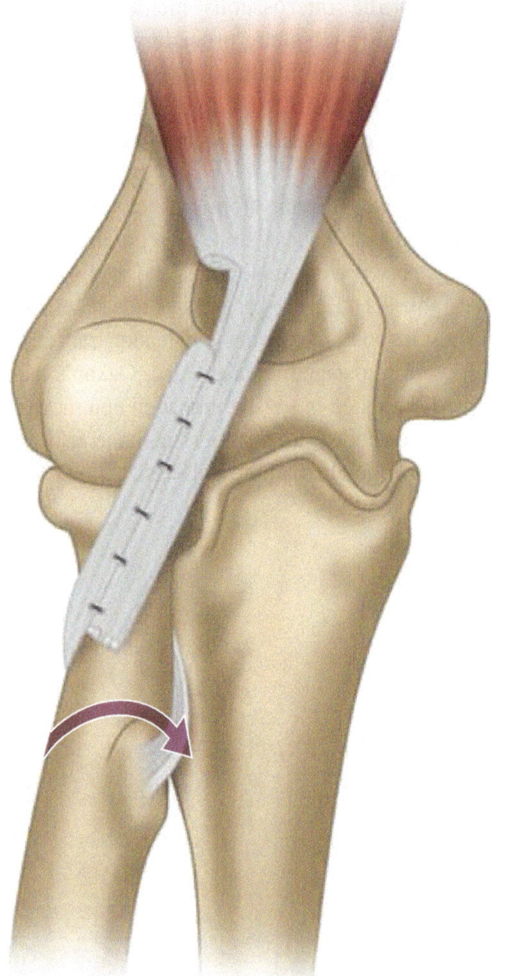

Fig. 7.10 Zancolli Grilli procedure step 2

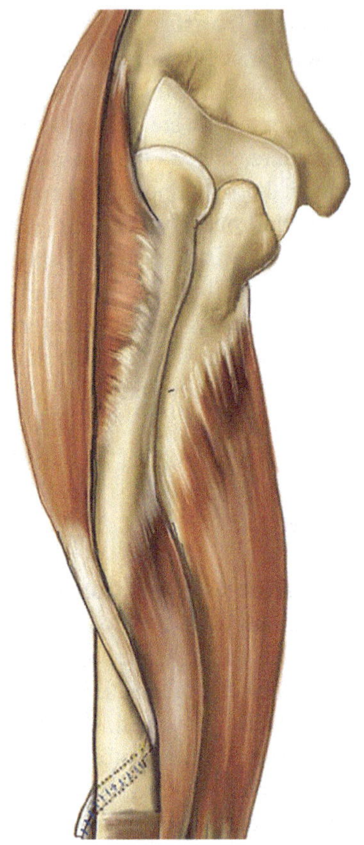

Fig. 7.11 Özkan Brachioradialis tendon rerouting

Table 7.1 Classification of forearm pronation contractures according to Gschwind

1. Active supination beyond neutral position
2. Active supination at or below neutral position
3. No active, but well-passive supination
4. No active, and limited passive supination

7.3.3 Wrist

7.3.3.1 Reanimation of Wrist Extension

Here, we routinely perform a tendon transfer onto the ECRB—but we are aware that the muscle donor quality is difficult to be assessed preoperatively (FCU or PT).

Sometimes, there is a separate or additional need to enhance the EDC (by one of the two donors).

7.3.3.2 Correction of Ulnar Deviation of the Wrist Deformity (UDWD)

The main finding of a study we conducted recently [8] was that ulnar head dysplasia, with hypoplastic deformity or overgrowth of the distal ulna, a shortened ulnar diaphysis, and/or an incongruent DRUJ, was identified in 18 patients with UDWD, who had completed radiological examination. The prevalence of the condition in our cohort of treated OBPP patients was approximately 2%. Despite this, the cohort was selected, consisting mainly of secondary referrals.

To date, there is neither an evidence-based treatment for UDWD nor a surgical strategy to reconstruct both the shortened ulna in a growing child and a functional DRUJ, especially as muscle weakness in the wrist stabilizers and the forearm pronators/supinators is frequently present. Decreased elbow function, by extension deficit and a commonly occurring restriction in shoulder external rotation, together with individual variations in long-term outcome, suggests that ongoing follow-up of children with OBPP is required beyond their preschool years. The clinical sign of uncontrolled ulnar wrist deviation in a child with an extensive obstetric palsy might be the visible epiphenomenon of a complex developmental disturbance of the DRUJ and its active motors and stabilizers. The severity of the initial OBPP related to ulnar nerve lesion (in particular total OBPP lesions) seems to be a common and determining pathophysiologic factor, in terms of the severity of the clinical findings of ulnar dysplasia and the radiological changes. Treatment should focus on the reconstruction of a rotational balance in prosupination and increased muscle strength. Ulnar deviation deformity of the wrist is not uncommon in children with incomplete recovery and is associated with permanent limitations in the use of their affected wrist and hand. The deforming force is most often an unbalanced ECU and/or FCU. Transfer of the deforming force to a more central location, usually to extensor carpi radialis brevis, is a common treatment, which to some extent could improve function and cosmetic appearance. The FCU to rerouted extensor carpi radialis longus (ECRL), the ECU to rerouted ECRL, the ECU to rerouted abductor pollicis longus, and the ECRL can also be surgical techniques to correct wrist ulnar deviation, but our results are, so far, unreliable. To centralize the ulnar head against the sigmoid notch of the radius, and thus improve the congruency of the DRUJ, earlier biceps rerouting and forearm rotation osteotomies may be recommended. The continued normal evolution and development of a normal DRUJ are dependent on the load and pressure given when the ulnar head is in its normal localization. Surgical procedures to lengthen the shortened ulnar diaphysis or to realign the DRUJ are so far hypothetical in a growing child. The IOM could also be addressed surgically, both to improve the passive ROM of prosupination, especially in the cases of supination deformity, and theoretically to create additional TFCC stabilizers in older patients.

We observed different modalities in terms of delayed skeletal ulnar growth and incongruence or dysplasia of the DRUJ. Unfortunately, the retrospective nature of this investigation and the limited size of our cohort did not allow establishing correlations between the UDWD radiological patterns and type of OBPP injury. Limitations of this study are the retrospective analysis, incomplete clinical data, limited and not standardized radiographs or equivalent investigation, and actual lack of clear treatment strategy in growing children. Another limitation is the selected cohort, consisting mainly of secondary referrals. However, the UDWD in patients with OBPP is, to our knowledge, not described in the literature before.

A follow-up of the patients operated on with biceps rerouting and rotational osteotomies, in terms of evaluating the potential correction of the prior ulnar hypoplasia, would be an interesting issue to examine further, particularly in terms of the possibility to gain a normal DRUJ appearance. Furthermore, a more detailed, prospective investigation with more patients, most likely through multicenter studies, and systematic extensive diagnosis and functional evaluation would be needed to be able to propose a classification system according to the severity of radiological findings and initial OBPP injury pattern. Such a classification—which was not possible to propose in terms of our case series cohort—could be adapted to different ages and injury patterns and would support and possibly clarify the choice of corrective surgery in these children with UDWD.

If the standardized wrist X-ray shows evidence of ulnar variance, what could be the potential corrections in the still growing child?

7.3.4 Hand

Proper hand function may be analyzed according to multiple concepts.

In the field of reconstructive hand surgery, especially when dealing with congenital variants, there is the concept of a "basic hand" [9] and its specific sensitive and motor requirements. General hand surgeons learn to deal with sequelae of median, ulnar, and radial nerve palsies.

My first group of patients with really impaired hands are the children with sequelae from total obstetric brachial plexus palsy.

While in the upper lesions affecting the C5 and C6 roots, the hand function is regularly good (although the median nerve might be impaired, thus affecting grip strength and sensation); as soon as C7 and C8 are involved, the wrist drops, and in severe total palsies, accompanied by C8 and Th1 root avulsions, the fingers become flail and insensate.

It was Alain Gilbert's major achievement to focus on the microsurgical reconstruction of the lower trunk in total obstetric palsies, in contrary to the current strategy in adults. Thereby, a lot of lost hands returned to a certainly very basic, but really functional activity.

Raimondi [10] classified hand function in obstetric palsy, and this was the first frame to analyze and score (Table 7.2).

Table 7.2 Raimondi classification of hand function in severe OBPP

Raimondi hand score	
Grade	Clinical finding
0.	Complete paralysis or functionally useless finger flexion
	Non-usable thumbs without grasping function
	Little or no sensation
1.	Limited finger flexion
	No finger or wrist extension
	Key grip possible
2.	Active wrist extension
	Passive flexion of fingers (tenodesis)
	Passive key grip in pronation
3.	Complete active finger and wrist flexion
	Active thumb movement, including abduction and opposition
	Intrinsic equilibrium
	No active supination
4.	Complete active finger and wrist flexion
	Active wrist extension, but weak finger extension
	Good opposition of thumb with active ulnar intrinsic muscles
	Partial pronation and supination
5.	Grade 4, but with active finger extension
	Complete pronation and supination

In total obstetric palsy, repaired or not, the recovered hand is smaller, without intrinsic function, with a basic grip and weak finger extension. It is and always will be a helping, assisting hand.

In cerebral palsy, spasticity adds a contracture of the first web space and an adducted, flexed thumb—altogether with flexed fingers.

Here, the treatment must address separately the inconvenience of uncontrollable spastic hypertonia and address the correct positioning of forearm and wrist, the use of the tenodesis effect, and a primary global fisting.

In tetraplegic patients, the goals for a basic hand function are either a passive or even better an active key pinch (lateral pinch) and an improved global long finger flexion, by empowering of the finger flexor muscles.

Disbalance between flexor and extending forces may lead to long finger deformities (especially swan neck), and even more complications through spastic hypertonia or when the intrinsic muscles of the hand, nearly always lost in peripheral nerve injuries, get tight due to spastic intrinsic involvement.

These are the global patterns we see, and it is then of utmost importance to enquire about the patient's intellectual capacities (affected in cerebral palsy or stroke) and his/her motivation. A hand without basic sensation will rarely be integrated, and the motivation for concrete tasks will be the motor for a reconstructive plan.

I choose to describe possible surgical strategies according to the various functional goals, not related to the underlying peripheral or central nerve lesion pattern, which of course will influence the type of surgery.

7.3.4.1 Improvement of Finger Flexion

This is done by adding functional muscle mass by the transfer of a pedicled muscle (FCR, LD, brachialis) or a free functional muscle transfer [11] (gracilis muscle, like the second step in

Doi's concept [12] for late complete adult brachial plexus lesions).

7.3.4.2 Improvement of Finger Extension

No individual finger extension may be achieved.

For MP joint extension, we consider a PT or FCU tendon transfer onto the fused EDC tendons.

For PIP and DIP extension, similar tendon transfers with long interposition tendon grafts (palmaris longus as a donor) onto the distal extensor apparatus (lateral bands) may be proposed in very selected cases (risk of long adherences, difficult reeducation).

7.3.4.3 MP Extension Contracture

Most important is prevention by casting in MP flexion-intrinsic plus position.

Dorsal capsulotomy and arthrolysis may help, followed by immediate and continuous passive mobilization after the surgery.

7.3.4.4 Late Basic Hand Reanimation

For sensation: intercostobrachialis (ICB, second intercostal sensitive nerve) to medianus nerve (lateral part) transfer.

Free gracilis for basic grip (Sect. 7.3.4.2), but problem of long-standing neglect of the very severely affected extremity, especially if sensation of the hand is poor or absent.

References

1. Kwong S, Kothary S, Poncinelli LL. Skeletal development of the proximal humerus in the pediatric population: MRI features. Am J Radiol. 2014;202: 418–25.
2. Birch R. Medial rotation contracture and posterior dislocation of the shoulder. In: Gilbert A, editor. Brachial plexus injuries. London: Dunitz; 2001.
3. Waters PM, Smith GR, Jaramillo D. Glenohumeral deformity secondary to brachial plexus birth palsy. J Bone Jt Surg. 1998;80A:668–77.
4. Bahm J. Changes in rotatory movements of the shoulder after obstetric brachial plexus lesion: clinical condition, surgery, and analysis of objective prognostic factors thesis. Brussels: ULB Brussels; 2011.
5. Waters PM, Bae DS. Effect of tendon transfers and extra-articular soft-tissue balancing on glenohumeral development in brachial plexus birth palsy. J Bone Jt Surg. 2005;87A:320–5.
6. Gschwind C, Tonkin M. Surgery for cerebral palsy: part 1: classification and operative procedures for pronation deformity. J Hand Surg Br. 1992;17:391–5.
7. Özkan T, Aydin HU, Berköz Ö, Özkan S, Kozanoğlu E. 'Switch' technique to restore pronation and radial deviation in 17 patients with brachial plexus birth palsy. J Hand Surg Eur Vol. 2019;44(9):905–12.
8. Bahm J, Bouslama S, Hagert EM, Andersson JK. Ulnar wrist deviation in children with Obstetric Brachial Plexus Palsy: A Descriptive Study of Clinical and Radiological Findings of Impaired Ulnar Growth and Associated Incongruence of the Distal Radioulnar Joint. Hand (N Y). 2020;15(5):615–9.
9. Entin MA. Salvaging the basic hand. Surg Clin N Am. 1968;48(5):1063–81.
10. Raimondi P. Evaluation of results in obstetric brachial plexus palsy: the hand. In: Presented at the international meeting on obstetric brachial plexus palsy. Heerlen: OBPI; 1993.
11. Bahm J, Ocampo-Pavcz C. Free functional gracilis muscle transfer in children with severe sequelae from obstetric brachial plexus palsy. J Brach Plex Periph Nerve Inj. 2008;3:23. https://doi.org/10.1186/1749-7221-3-23.
12. Doi K, Hattori Y, Sakamoto S, Dodakundi C, Satbhai NG, Montales T. Current procedure of double free muscle transfer for traumatic total brachial plexus palsy. JBJS Essent Surg Tech. 2013;3(3):e16.

8 Postoperative Rehabilitation After Reconstructive Surgery: Interaction with Physio- and Occupational Therapists

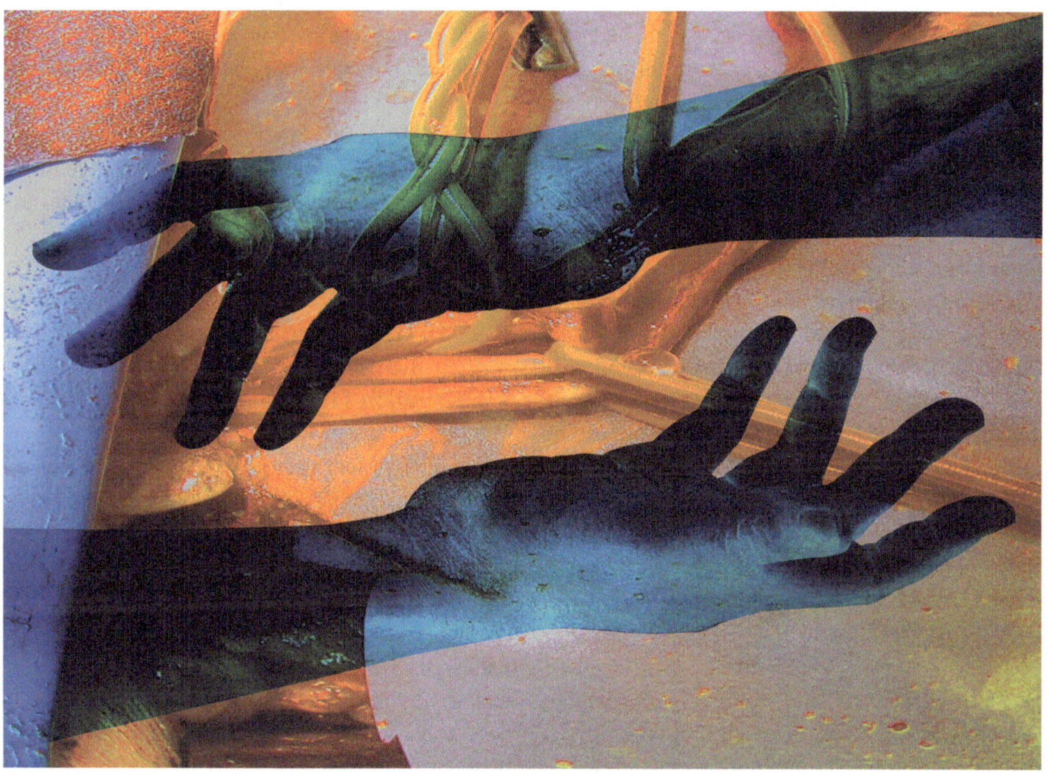

J. Bahm, *Surgical Rationales in Functional Reconstructive Surgery of the Upper Extremity*,
https://doi.org/10.1007/978-3-031-32005-7_8

Recovery of motor or sensory function after reconstructive surgery does not happen just by waiting. Specialized therapists support and accompany the biologic healing processes, like wound healing and nerve regeneration, bone healing, and muscle reinnervation.

It is therefore mandatory to establish and maintain a regular dialogue with the concerned physio- and ergotherapists, to understand their education and motivation, and to communicate what we operated on and what we expect in a specific time frame. We are not always aware of their techniques and scientific background, and therefore a common base has to be defined (we call it the "Common Pathophysiological Basis"; see 8.3). As their education is established in a program which is independent from our medical education, the common language may be based on anatomy and biology, and derived pathophysiology, but with the ongoing development of further concepts like in osteopathy or fascial techniques, we as surgeons also should open our mind for new theories and approaches.

In some countries, the patients themselves redefine whom they address for first-line health care—for a lot of functional purposes and pain issues concerning the musculoskeletal system, they consult first the osteopath before searching medical advice.

8.1 How the Surgeons Consider Postoperative Recovery of Function

Our aim is recovery and improvement of pain-free passive and active range of motion, within a normal sensorium, allowing the patient to be more performant in activities of daily living, sports, or leisure activities.

Our surgery is thus focused on increasing joint mobility, strengthening motor function, improving sensation and reducing pain, and looking at the achievement of particular tasks with the addressed limb. Skin and soft tissue wound healing requires 2 weeks of reduced mobilization, tendon sutures regularly, and 6 weeks of rest. Nerve coaptation is resistant after 2 weeks. Bone healing may last for 2–3 months.

This knowledge directs our recommendations concerning the onset of passive and active mobilization in a therapeutic setting.

We fear soft tissue adhesions affecting the gliding planes of tendons and therefore recommend "early" active mobilization whenever possible. Recovery of motor strength is the long-term result of regular use under slowly increasing demand (following the BMRC grading).

We might tend to render the therapist responsible for the quality of progress in revalidation and the endpoint reached—frequently placing his role above the limited capacities and compliance of the patient. The final outcome is supposed to mark the success of our surgical work and will influence our rationale behind indications for surgery.

With ongoing practice, we become more aware of this complex interaction and feedback loop and are more able to integrate relevant outcome factors.

8.2 The Interdisciplinary Continuum: Talking with Therapists

This is the perfect exercise of a dialogue between different disciplines. Although the administrative regulations bias the leveling between the prescribing doctor and the executing physio- or ergotherapist, there is a daily reality where the therapist in his relation with the patient establishes his own treatment plan and applies his techniques in order to achieve a functional goal acceptable for the patient, the surgeon, and finally the therapist himself.

This is the challenge of interdisciplinarity, and the dialogue is the constructive challenge to make it happen. It should be respectful, regular, informative, and open minded. Practically, it is mediated by the telephone or an internet-based video conference (Teams, Zoom), and in some centers, a physical meeting and direct exchange on the ward or in the therapeutic setting are feasible.

A common pathophysiological base is helpful, but it is not the only field of exchange. If we are able to listen to other thoughts and techniques, we will learn about this other discipline and thus create more confidence-based links. The patient will feel more secure and happy in an environment where the doctor and therapist interact regularly.

Regular exchange may allow stepwise adaptation of the techniques and progressive functional follow-up. New techniques might be added (e.g., ergotherapy), and the focus may be rearranged.

It is not necessarily time consuming, and regular, e.g., weekly, staff meetings will produce an effect of continuous education.

8.3 Common Pathophysiological Basis

From the medical point of view, the biologic pathophysiology of wound healing, extended to bone and tendon healing, altogether with basic knowledge on peripheral nerve de- and regeneration, muscle atrophy and sensory recovery (especially proprioception) will set a good base.

Pain issues are also important, especially neuropathic pain and common understanding of the complex regional pain syndrome (CRPS). Independent of origin and specific pathophysiology, pain may always be assessed and monitored using the visual analogue scale (VAS).

8.3.1 Proprioception

Reeducation of ankle ligament sprains is usually based on proprioceptive concepts. Recently, the same principles have been investigated for the upper limb and especially the wrist [1] and thereby shed new light on the importance of joint innervation (by the posterior interosseous nerve) and the stabilizing role of tendons on the moving wrist (a concept well indicated in young women suffering from painful wrists with ligament hyperlaxity, where surgery is contraindicated).

As ergotherapy becomes more regular in hand surgery departments, we do hope that these specific reeducation skills will progressively enter routine treatment protocols and also become part of the arsenal of therapists dedicated to our upper limb reconstructive surgery.

8.4 Where Do We Need Improvement

8.4.1 Timing

Babies or young children affected by any motion disorder are very rapidly directed towards physiotherapy, a logical and mandatory step. Depending on their diagnosis, this conservative treatment is provided for years. In my experience with children affected by a traumatic condition, i.e., obstetric brachial plexus palsy, it is difficult to set an endpoint, because parents always have the feeling one wants to withdraw something helpful, perhaps perceived as essential, and therapists prepare the parents over time to the nearly never-ending support to prevent bad posture, increased neglect, and less skills. In a terrible situation with an economic background as in our country, physiotherapy is provided on medical prescription and is budgeted restrictively for most of the health institutions.

My analysis starts with the need of a precise *functional goal* for any type of conservative revalidation, and as long as there is a clear goal which may be reasonably reached, I fight with the parents and therapists to maintain that situation.

Only in young and preschool kids do I consider also the preventive character of treatment; further on, I advise sport activities, exploring different types, especially swimming (global muscle activity, body symmetry) and horse riding (axial alignment, self-confidence). Older children should be educated and encouraged to practice physical activity regularly and to have fun with those activities. They need to test some sports and find out by themselves what they really like.

All parents accept to stop physiotherapy once the child starts to resist, but they feel that our intervention (based on sound medical arguments) only serves the “system” and is harmful for their children—strange.

8.4.2 Type of Therapy

Also, the type must be adapted to the child's or patient's condition. The so-called neurophysiology-based therapies like Vojta and even Bobath apply only as long as the neuroregenerative process is working—and this means in obstetric palsy 2–3 years, not longer. Ergotherapy with exercises focusing on daily activities may be a good follower for compliant and sufficiently able children.

In all patients, postoperative therapy is surgery specific and should be of course adapted to the patient and the precise recommendations of the surgical team. When transferring a tonic-type muscle like the latissimus dorsi into a new function which requires phasic activity, a specific training program may last 12–18 months and must consider that the morphology of the initially tonic muscle with its specific muscle fiber types must change to new fiber profiles adapted to the new, fast-twitch activity—here, regular therapy is mandatory throughout the whole morphologic and physiologic adaptation process.

Unfortunately, most prescribing physicians do not know or do not show interest for physiotherapist's specific education and skills. Thereby, although the initial signal and the administrative support come from the doctor, the physiotherapist is mostly left alone in his daily practice, assuming at his best the revalidation work, without any feedback to the prescribing doctor.

8.4.3 Interaction

Since the beginning of my career, I got used to phone very regularly to the therapists, to get my work known, to inquire about the way they treat children and adults, to ask for the observed progress, and to adapt the postoperative training to my specific considerations, derived from intraoperative observation (i.e., muscle quality) or specific technical characteristics (tendon-bone anchoring by specific anchors using nonabsorbable, wired threats which resist much earlier to training). This always has been very rewarding, and I remember having visited several specialized training centers, observing their work with all age patients, and thereby learning myself much more about muscle physiology, training programs, and overall the potentials and also scientific and pedagogic background of their therapies.

In my consultant activity at ULB Erasme University Hospital in Brussels, my consultation always hosts one physiotherapist and several students from the Faculty of Motor Sciences (studying either physiotherapy or osteopathy). This interdisciplinary approach has proven to be very fruitful over the last 15 years. Many students choose to write their "thesis" on specific aspects of our treatment protocols, and this is definitely a typical win–win.

8.5 Sensation

We still feel that sensory reeducation remains as the orphan part, although a lot of research and written information have been accomplished (Lundborg 2000; Spicher 2003–2020; Dellon 2019; Moseley).

One problem might be the individual and complex character of sensation and pain issues, and the other the long-lasting therapy and the concomitant diseases (depression, fibromyalgia, chronic pain).

Sensation is easy to be assessed and monitored, and there are actually valid therapies like the graded motor imagery program (Moseley) and sensitive reeducation of neuropathic pain (Spicher).

Cortical (and more precisely) neuronal plasticity is an exciting field of research with rising clinical impact, giving hope to a lot of difficult issues in all age groups.

Surgery to improve upper limb pain or to restore sensation is a still evolving field and is summarized in "Pain Issues".

8.6 Pain Issues

Upper limb pain for other reasons than mechanical or inflammatory origin becomes more frequent and difficult to treat. Neuropathic origin is suspected when the description characterizes a shooting pain, going along nerve axes, felt like burning or severe tissue compression, and when common diagnoses of arthrosis, tendinitis, and overuse have been ruled out.

Severe and chronic neuropathic pain may be associated with CRPS type II (complex regional pain syndrome) including an underlying nerve fiber lesion, dystrophic changes of edema, allodynia or hyperesthesia, and functional loss like in the classic Sudeck's algodystrophy.

A complete neurologic assessment is mandatory, including imagery, starting at the cervical spine, looking at the myelon, intervertebral discs, foraminal or shoulder impingement, and arm and forearm compression sites for peripheral nerves. Inflammatory or viral neuropathies, especially Lyme's disease, must be excluded.

8.6.1 Neuropathic Pain: Interdisciplinarity

Treatment of neuropathic pain challenges neurologists, anesthesiologists, and orthopedic and hand surgeons, but also more neurosurgeons involved with invasive neurostimulation (neuromodulation).

The first line includes drugs having neuroleptic (pregabalin), antiepileptic (carbamazepine), and antidepressive properties, with associated physiotherapy.

Conditions with surgical solutions must be actively looked for and eventually be treated by surgical means (peripheral nerve decompression, thoracic outlet, narrow medullary canal, or disc protrusion).

But neuropathic pain frequently gets chronic and invades all moments of daily living activities—so, physiotherapy, behavior psychotherapy, and occupational therapy are as important as medication. The main goal is to bring the patient back to a regular life ("participation"), and there are many strategies to try this, even a complete withdrawal of all medication (like for drug and alcohol abuse), resulting in a physical and psychological deprivation to be treated under hospital conditions, associated with continuous occupational treatment.

The actual neurosurgical tendency is invasive neurostimulation at different levels (**neuromodulation**). Based on the experience with TENS therapy, and the positive effect of electrical stimulation on sensitive nerves responsible for local superficial pain, the actual electrodes are no longer epicutaneous, but placed more proximally on the proximal nerves, medullary ganglions, or even intracerebral relay stations like the thalamus. Once the percutaneous stimulation has given sufficient relief, the stimulator and the electrodes are implanted, the latter close to the described target zones at peripheral, medullary, or cerebral level. The "pacemaker" is inserted in a subcutaneous pocket and is thus accessible for (re)programming and battery change.

Peripheral nerve surgeons, especially dedicated hand surgeons, developed other approaches to that complex problem, looking for peripheral nerve damage to be treated by direct microsurgery, developed in the following paragraph.

8.6.2 Peripheral Nerve Surgery for Neuropathic Pain Issues

If there is a chance that the neuropathic pain is caused or triggered by a peripheral (sensitive) nerve lesion, the first step is to identify and address this lesion, described as a neuroma in continuity or a stump neuroma.

The neuroma of a bigger nerve may be visualized by ultrasound and MRI and a test with infiltration of local anesthetic is performed. If the result on pain relief is satisfactory, not only does it provide the patient with a feeling of how the exclusion of this nerve-driven information will improve the symptoms, but it also indicates the surgeon the place to investigate further.

On local exploration, the affected nerve may be found adherent and scarred (and after neurolysis, one must prepare a healthy vascularized soft tissue bed to embed the freed nerve segment—otherwise, short-term recurrence of the painful symptoms is guaranteed) and one may find a lesioned area, either fibrotic or elaborated as a neuroma, in continuity of the nerve or as an end structure (stump neuroma).

The neuroma is made of the disordered assembly of free nerve endings, which in their regenerative impulsion either tried to cross a gap (and succeeded more or less) or tried to find the distal pathway which turned out to be never accessible (stump neuroma).

Any excision of the neuroma minifascicles will reproduce a fresh traumatic condition and reinitialize nerve fiber regeneration, thus reforming a neuroma (recurrence). The actual strategy is thus as follows:

- In a neuroma in continuity, to resect the lesioned part and interpose an autograft, so to transform a situation with difficult regenerative condition in a more favorable one.
- In a stump neuroma: either to accept the local bundle of triggering minifascicles and to hide it from surface contact (bearing it into a deep muscle, or bone cavity—with no more access to the superficial soft tissue or skin layer) or to definitely target the recut stump with a sensible end organ (skin—a piece of deepithelized dermis, wrapped around the nerve end). Attempts to re-suture the free nerve ending onto itself (omega suture or end to side) only follow a morphological argument of tubing but not the physiology of regeneration and thus are rather abandoned.

The dermal cup is the easiest way of retargeting sensible fibers, and with the actual development of nerve transfers, more elaborated ways of attribution of new sensitive targets may be developed, using sensible nerve transfers.

In this novel area, I got surprised about our poor knowledge on sensitive organs, their behavior under the condition of nerve fiber degeneration, and the possibility of restoring sensation by specific sensitive training.

8.6.3 Neurogenic and Disputed Thoracic Outlet: Differential Diagnosis and Surgery

Peripheral nerve compression in the upper limb happens at various specific anatomical landmarks, with different frequencies: compression of the median nerve at the carpal tunnel and of the ulnar nerve at the elbow are the most frequent compressive neuropathies of the upper limb; the pronator syndrome (median nerve compression at the origin of the pronator muscle at the volar elbow) and the compression of the ulnar nerve at Guyon's canal are much rarer. The radial nerve may be irritated under the supinator arch, and all upper limb nerves, especially those deriving from the medial cord, may get irritated at the so-called thoracic outlet region.

Vascular TOS problems are regularly addressed to the vascular surgeon and thereby rarely seen in our consultation; but all those with issues of pain and weakness are addressed and need a careful clinical and neurological workup.

If the neurologic testing, including dynamic MR imagery and electrophysiologic examination, finds arguments for nerve compression, especially on the lower trunk/medial cord, a neuro-TOS is suspected and indicated for surgery, once a conservative treatment trial for 3–6 months under good conditions has failed.

The other cases are called "disputed," as there are neither neurologic nor vascular arguments for compression, but still impressive symptoms. The surgical indication is very delicate, as one wants to help the patient, but also to avoid postoperative complications, especially enhanced neurologic symptoms, limb apraxia, and major sensitive complaints. When all paraclinical exams are normal (most of them are done under static conditions), when the conservative treatment failed, and when the patient has already seen many colleagues with finally insufficient conclusions, then the difficult decision about a "surgical diagnostic exploration" has to be done.

My actual strategy is a repetitive, very elaborated open-minded discussion with the patient, going through all scenarios, including a possible difficult postoperative course. The decision is never taken at the first consultation, and the patient is addressed in a way that he clearly takes the final decision, being informed about all positive and negative issues for his limb and life.

Interestingly, surgical exploration in this group still finds minor or major anatomic variations (mostly muscle variants like absence of the interscalene gap, hypertrophied anterior scalenus muscle, existence of a scalenus minimus), and sometimes a lasting postoperative improvement without clear intraoperative correlate. In rare cases, the costoclavicular narrowing during abduction and external rotation of the shoulder compresses the lower trunk like a bony scissoring and needs resection of the first rib.

8.6.4 Conservative Treatment of TOS

Peet [2] proposed the term thoracic outlet and introduced a first scheme of conservative approach; he has been successively cited by a lot of authors like Roos and Wilhelm.

Vanti et al. [3] proposed a review and analysis of evidence, addressing concepts like improvement of posture, activation of muscles which could open the costoclavicular space, relaxing those responsible for a closure, adaptation to activities of daily living, and pain compensation, but in summary, besides a table with recommendations, there is nothing very conclusive or specific.

Most surgeons do not even know what could be the good principles of a conservative approach!

Approach in osteopathy should include analysis of functional disturbance of the musculoskeletal system, i.e., muscle imbalance of the cervical spine and shoulder girdle; a bad posture; weakness; and overuse hypertrophy of certain muscles. Patient's compliance to that treatment before surgery is very important; therefore, it is necessary to refer to an informed and skilled therapist and to maintain a continuous dialogue.

8.6.5 De-escalation of Surgery for Neuropathic Pain Issues

My experience with patients suffering from chronic neuropathic pain of the upper limb is that of recurrent surgery and increase in invasive neurosurgical procedures, summarized under the rather smart term of "neuromodulation."

The first setting is a kind of medical shopping among surgical specialties (orthopedics, hand surgeons) with sometimes recurrent decompressive surgery, followed by longer periods of physiotherapy without restoring the working capacity. We see frequently young patients, out of work for two or more years or sent to university to rule out a TOS or rare neurologic conditions—at a moment, the pain memory has already been fully activated.

Once the "pain team" made of anesthetists resigns, neurosurgeons are the next step, proposing different peripheral or more central external or implantable stimulation devices, culminating in implanted para-radicular electrodes and even open cerebral surgery involving the thalamus.

We met recently a neuro-orthopedic colleague promoting a totally different but also radical approach, where the patient undergoes a 3-week withdrawal of all morphinic drugs altogether with daily integrative exercises—here, participation in usual daily activities including work scenarios is the mantra. This is an uncommon and long way, but probably so closer to the real aim of our active life, and apparently effective also in chronic cases with proven pain memory.

References

1. Hagert E. Proprioception of the wrist joint: a review of current concepts and possible implications on the rehabilitation of the wrist. J Hand Ther. 2010;23(1):2–17.
2. Peet RM, Henriksen JD, Anderson TP, Martin GM. Thoracic-outlet syndrome: evaluation of a therapeutic exercise program. Proc Staff Meet Mayo Clin. 1956;31(9):281–7.
3. Vanti C, Natalini L, Romeo A, Tosarelli D, Pillastrini P. Conservative treatment of thoracic outlet syndrome. A review of the literature. Eura Medicophys. 2007;43(1):55–70.

9 Dialogue Between Surgical Clinics and Research: How Could We Share Projects and Strategic Directions?

J. Bahm, *Surgical Rationales in Functional Reconstructive Surgery of the Upper Extremity*,
https://doi.org/10.1007/978-3-031-32005-7_9

In the last 25 years, I had the chance to follow lots of congresses dedicated to peripheral nerve and brachial plexus surgery. I distinguished two types of sessions: those relating to surgical technique and postoperative results and those addressing basic science.

The most frequent basic science subject has been and probably still is the search for substitutes to autologous nerve grafts.

Generally, the presenters start pointing a clinical need for more than the available sural and saphenous nerve grafts, and eventually the harm when harvesting autologous nerves (donor defect).

In my clinical experience, I did not encounter one situation where I clearly missed more grafts. This may depend on my activity and the country I work (I may imagine that for long sciatic nerve defects we may run short of autologous grafts, but I have no experience with those war-related injuries). I also did not see major issues with donor defects after sural nerve harvest, the most frequent donor nerve for autologous grafts.

This does not mean that I see no rationale for developing bioartificial nerve grafts, tubes, and processed allografts—but the rationale behind should be adapted to the clinical reality and there should be much more dialogue between searchers and surgeons.

Nerve regeneration is a fascinating biological process, and it is absolutely necessary for everybody who deals with peripheral nerve reconstruction to understand this healing factory and to promote or accompany search for more insight. I believe that biologists dedicated to lab work are much more prepared and skilled to investigate this subject further, as they learn already during their studies much more in that field than a future medical doctor will ever do. Also, they do have the advantage to be able to compare this process in different animals; they know animals much better considering their anatomy, physiology, evolution … than we ever will.

During a medical career, scientific work may have an important place for those who plan an academic career, and the so-called experimental work is valuated for both MD and PhD degrees—but it will never transform a doctor into a brilliant scientist (or a brilliant scientist in a good doctor or surgeon … per se). But the rules and common agreements exist; no singular person will ever change the "system."

My own philosophy has always been that scientific effort should be related to clinical improvement, and so, my PhD thesis treated the understanding and treatment of the rotational imbalance of the glenohumeral joint after obstetric palsy. What I discovered through the biomechanical approach helped me to re-evaluate and improve the different surgeries I apply to rebalance this joint. But I agree that this is only one and very personal attitude.

I also understand that those investing their energy in basic science aimed to understand important biological processes or to help setting a basis for fundamental knowledge, like the vast area of the so-called tissue engineering. Even if my younger colleagues will participate part-time for 2 or 3 years in a laboratory experiment, neither before nor after this period they will be full-time biologists and their future medical practice will not be influenced in the long term by those skills and results they got in the lab. They will be able to understand other searchers, and they will understand related papers, but this will not necessarily frame their future evolution as a clinician or surgeon—as even the microsurgery we perform stops at a level of fivefold magnification and will never dive into the microcosms of cell populations, like during any search activity in bioengineering science.

Looking with a wider perspective because of coming close to a generation time span, I notice that, e.g., the promises given some 20 years ago by the enthusiastic starters in bioengineering (replace like with like) do not stand today with clear achievements—we have no solution for major cartilage loss, for composite tissue defects, and for functional tissue replacement (neuromuscular units). This is not an argument to stop here, but just to allow a discussion about what should be the priority, what are the real challenges, and who is "allowed" to define them.

The actual pandemia of coronavirus infection redefines a lot of needs in our health system and in our basic understanding of viral infection, immunity, vaccination, and animal–human interaction. Here, the definition of priorities is conducted by emergency and a life-threatening pandemia. All aspects of social life are subordinated to that condition.

Health conditions in a lot of economically less powerful countries are no longer priorities—we just do not even talk or think about it.

Still, all children in the world are our common future, and everything done for children's health should be a matter of attention. Is it so? In our Western countries, medical research and granting for projects involving children are not a priority. In these times, much more is done for the elderly population, the research around dementia, and the establishment of proper institutions for the dependent older citizens—a certainly respectable strategy, but a priority driven by sometimes unclear motivation.

Who has the right and duty to decide? Where is the searcher's own free will? How may we adapt and integrate different views? Where happens the public debate? Over the years, it has always frightened me to see how poor is our patients' knowledge about the real structure and economy of the national health system, its priorities, and its decision pathways. I always encourage them to get more conscious and to influence decisions by their democratic vote. Now in corona times, everybody is concerned with his health, the well-being of his family and relatives, the good functioning of a health system, and the respect of all care professionals, especially the nurses. But how long this will last and how deep will further conscious development go?

Everybody in the great medical community has his own (protective) arguments, a rational self-defense. The need for funding in the scientific community is a daily reality, and the overall opinion of the population is driven by the focus given by major print and digital media.

Finally, we are all responsible, but are we sufficiently conscious, unbiased, and open minded?

These lines can only be a stimulus to go further, read more, and listen and talk to representatives from all interested groups, and this stimulus points to the constant need for compromises, honest statements, and constructive dialogues.

Perhaps, the medical community should dialogue much more with the searchers in the respective field, raising questions, defining needs, and highlighting misdirections. But this means also that we should learn more about the scientist's mind, their specific way of thinking, and their language. I remember that the exchanges at the scientific sessions I visited were always rather short, comprising an important percentage of misunderstanding and just parallel thinking, without finding intersection points. Probably, a lot of money and energy are wasted just because on both sides, colleagues do not bring the information and the project to that critical common point—which should finally serve the patient.

It must be terrible to understand after several years of serious investigation that the chosen way turns out to be wrong, misleading, and ineffective. It is also frustrating to deal with patients where basic science and research do not provide new insights, better help, and steps for a successful social integration (Table 9.1).

Table 9.1 Highlights several spots for further reading or further development

Think about:
1. Social integration of patients with upper limb disability—a major therapeutic goal
2. Lobby for OBPP kids in our society is very poor
3. Neural plasticity
4. Refine secondary surgery technique based on increased knowledge of their pathophysiology

10 Complications, Pitfalls, The Worst Case: How We Learn From Our Mistakes in Reconstructive Surgery

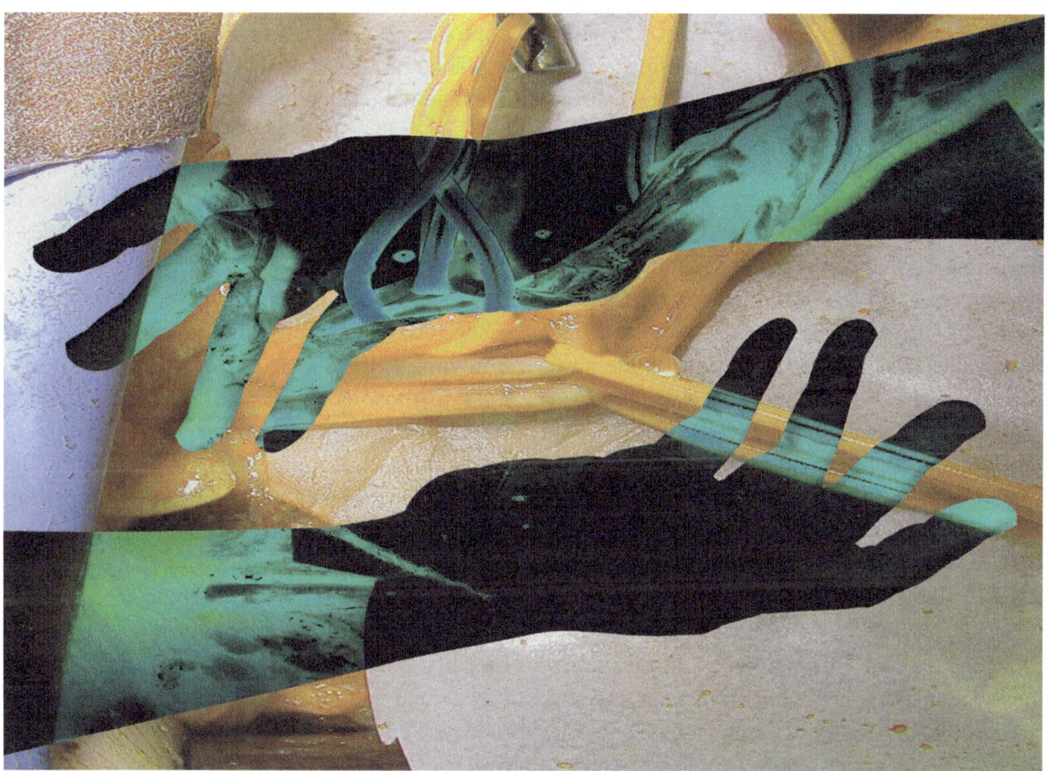

J. Bahm, *Surgical Rationales in Functional Reconstructive Surgery of the Upper Extremity*, https://doi.org/10.1007/978-3-031-32005-7_10

10.1 Failures of Free Flaps

Free flap surgery needs a solid team, where surgeons, operative nurses, and anesthesiologists work together in a constant and reliable setting.

Keeping the OR and the patient warm, maintaining fluid balance and blood tension at a constant level, and of course excluding coagulation disorders before any surgery are so easy but absolutely mandatory rules you must respect. I had to pay a tribute in situations where for various reasons I was unable to maintain those stable conditions, and it is a terrible experience for all participants, especially the patient and his family, but also for all members of the team, and here mostly the responsible surgeon. You feel abandoned, losing confidence. It is a moment where your enthusiasm stops, where you feel strong limits, and where you need help.

Therefore, when we actually do a functional free muscle flap transfer, we prefer a systematic two-team approach where the first team raises the flap and is further on responsible for the vascular anastomoses and the second team prepares the recipient site with the vessels and the donor nerve, fixes the muscle, and deals with the nerve coaptation. In this setting, the heavy burden is distributed on several shoulders, and the overall good outcome is checked by more than one surgeon.

10.2 Forensic Claims

Extended patient information is always absolutely mandatory, including a written consent, focused especially on complications and rare issues. For example, surgery aimed to relieve neuropathic pain could make the postoperative pain temporarily worse (I had this situation in a third surgery for recurrent TOS syndrome).

10.3 Concurrence Between Colleagues

Only a national concern (or even a regional one—when patient recruitment happens on similar fields) … rather difficult to talk about. We are all humans, but as such, we should have civilized relations especially with our peers. The patient should not suffer from disagreements among caregivers.

10.4 Postoperative Complications and Pain

When I was a last-year medical student, a lawyer gave us a lesson on how to deal with patients suffering from complications following our treatment or with those being unsatisfied. He gave the precious advice to spend a lot of energy staying very close to that patient until the situation improves. He urged us to provide all types of explanations, to stay listening, and thereby to alleviate the tension and the raising lack of trust. Over more than 30 years, while growing progressively into more and more personal responsibility in relation to the patients I took care, this lesson turned out to be one of the most precious rules I have ever learned.

We probably all know the reflex trying to escape from something difficult, disturbing, and troublesome. It is exactly that reflex we have to fight in those "complicated" relations, thereby maintaining open communication, developing empathy, and accompanying in friendship the patient who seeks to overcome a difficult period, especially after a surgical treatment.

Informed consent before any procedure is a good start, even better with a signed agreement—but we should not overestimate the patient's capacity to listen and to understand all we explain *before* and not underestimate his belief about riskless cure, magic medicine, and divine doctors.

Postoperative pain is very stressing and unpredictable. Surgery is finished, but the pain stays, and the patient has no signs which indicate improvement. In those moments, he needs us, our presence, reassurance, listening, and concrete action. It may be dangerous if this particular aspect is shifted only to the anesthesiologist or the ward nurse. We have to be present at the crucial moments, with clear words and decisions. Pain must be rapidly and adequately treated and the analgetic effect of the prescribed drugs regularly controlled—eventually with adaptation of amount and types of drugs delivered.

Other complications, like hematoma and infection, must be dealt with as much concern and professionalism as the primary surgery itself, even on weekends, holidays, and "inconvenient" occasions.

The time and energy we spend at those moments are like very-well-invested money, as the patient regularly will continue to trust us, accept drawbacks, stay away from legal issues, and show up for another surgery—as he feels that we are a constant support. Nobody is or needs to be perfect—just be present as a human friend and skilled surgeon.

There will be enough other moments where visits or explanations may be shortened (once everything goes well).

11 Continuity in Surgical Education

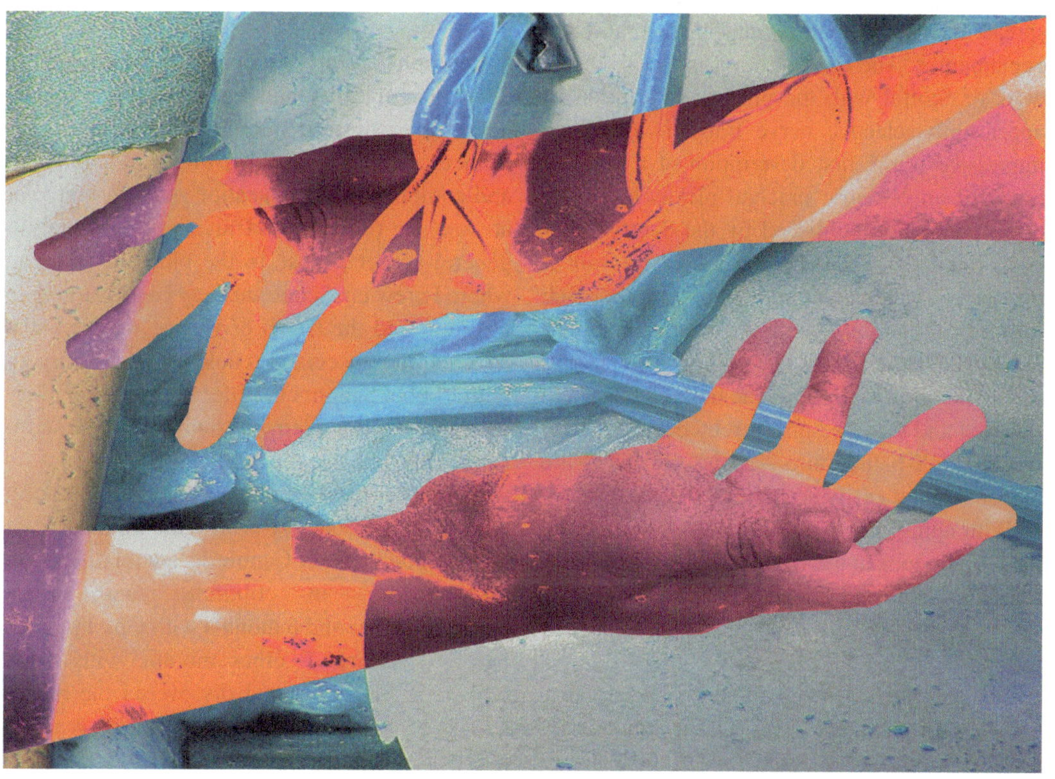

J. Bahm, *Surgical Rationales in Functional Reconstructive Surgery of the Upper Extremity*,
https://doi.org/10.1007/978-3-031-32005-7_11

11.1 How Historical Perspective May Shape Our Surgical Strategy

In our recent textbook [1], my colleague and friend Andreas Gohritz wrote an excellent chapter on the history of reconstructive surgery, especially peripheral nerve surgery.

The two world wars contributed a lot to the development of sound nerve surgery and so-called secondary, palliative procedures, as did the thrive to treat sequelae from poliomyelitis.

In the nineteenth and twentieth centuries, a lot of important papers were not written in English, but in German or French, and thus not widespread into all scientific and medico-surgical communities. The most striking is the extensive work of Stoffel with the precise description of upper limb selective nerve transfers [2]. Also impressive is the first description of a surgical repair of an obstetric brachial plexus lesion by direct coaptation published by the Scottish surgeon Kennedy [3]. This shows how relevant it may be to read ancient authors and to approach the medical literature including outstanding, non-English contributions.

But in our modern time, the first reflex goes to PubMed and science hub, thereby detecting only recent English literature—as the old gems are hidden and not easily accessible.

With aging, one may develop a more natural tendency to look at the work of others and to relativize our own achievements, coming back to the image where "we all stand on the shoulders of giants."

History may open our mind, teach modesty, and help for synthetic views. Old techniques are not always obsolete and may set a sound basis for new developments (like nerve transfers, see above). But we need enough time to read, to search, and to understand—sometimes to translate ... difficult in the more hasty and evidence-based time.

Historical contributions also add to artistic work (by their drawings) and cultural background of the various European societies. In that way, a regular exchange with dedicated colleagues from all over the world in congresses and further by emails may be very stimulating and constructive, as not only scientific aspects are debated further, but also cultural habits and viewpoints are elaborated.

I strongly wish to encourage every young and motivated colleague to travel, to meet the peers, and to discuss a lot—this will create long-lasting relations giving a much more "international" color to our daily work. It also allows distraction, new perspectives, and sometimes a genius moment of friendship and discovery.

11.2 Perspectives from an Individual Career

I accomplished my surgical education through different university and nonacademic hospitals in three countries: Germany, Belgium, and France: different cultures, different mentality, and complementary education. After a first year at Aachen (Germany) University Hospital in general surgery, close to the place in Belgium where I grew up, I continued my basic training in Belgium and the North of France, where I discovered the fascinating field of hand surgery with Michel Schoofs in Lille. I visited him every Tuesday as a complement to my surgical internship. This led me to apply for a residency in well-known hand surgery centers in France, allowing me to spend 1 year in the trauma center of Strasbourg (CTO Centre de Traumatologie et d'Orthopédie in Illkirch) with Jean Nonnenmacher and 6 months with Guy Foucher, before spending 5 other months with Alain Gilbert in Paris, where I discovered the reconstructive microsurgery of obstetric brachial plexus lesions.

I returned to Aachen at the end of 1994 to finish my general and plastic surgery residency at the same Aachen University Hospital where I had started in 1988, and I moved to a small general hospital in Aachen town in April 2000, to create my own unit of reconstructive microsurgery (and hand surgery) at the Franziskus Hospital, where I had the chance to develop my surgical activity and work with younger assistant surgeons and later on residents, until this hospital was finally integrated into the university hospital in the beginning of 2020.

This was the moment once again to reintegrate the same university hospital where I had started 32 years before, now as the head of a division for plexus surgery created within the department for plastic, hand, and burn surgery directed by Justus Beier.

Thus, the circle has closed, and for my last professional years, I may take advantage of all the connections this structure may offer and the huge technical and especially manpower support in anesthesia such a structure provides, altogether with the stimulating cooperation with young trainees, searchers, and medical students—some of them preparing their MD degree with various subjects related to the 2000 patient records of children affected by obstetric brachial plexus palsy treated in the Franziskus Hospital over the last 20 years (from 2000 to 2020).

So, I had the privilege to work in both nonacademic and university settings, the former offering good conditions for effective operation days—a lot of cases in 2–3 days a week—and the latter a safer and more modern background for the delicate nerve reconstruction in little babies, from 2 to 9 months old. Where I do more science and paperwork now, I was more in the OR in the previous setting, fulfilling my surgical expertise and developing new techniques (like the direct repair in selected upper and extended upper obstetric brachial plexus lesions) and new indications (nerve transfers in specific cases of atypical upper limb arthrogryposis).

Now, with the help of students willing to review the data of my surgical records, it is the time for evaluation, critical appreciation, and publication of results and indications.

In the actual environment, there is no more economic stress (a little hospital structure and administration make you feel the stress of regular income due to your operative activity), but the structure is more complex, the administration is much more elaborated, and the working processes are full of unpredictable details, innovations, and irregularities—due for example to the always changing co-workers in anesthesiology and operative nurses.

And now, the corona pandemia changed patient flows a lot, both at the outpatient clinic and OR activity, and the future still remains unknown.

My spectrum enlarged with time, allowing me to extend from obstetric brachial plexus cases to adults and to add children with arthrogryposis, cerebral palsy, and nowadays adults with hemi- or tetraplegia, with challenging innovations specially in the field of peripheral nerve surgery, and here the wider use of nerve transfers.

Also, I was able to stimulate younger research colleagues busy in tissue engineering of nerve and muscle cells, especially working on bioartificial nerve grafts, to get interested also in the fascinating activity of the neuromuscular junction or end plate, a field with a lot of implications in our motor reconstructive microsurgery.

Good connections with physiotherapists and pediatricians have always been an important basis for my patient recruitment, as extended availability for the parents of affected children and the (older) patients themselves. Before becoming a "specialist," I was just listening, at their disposal, curious, and willing to inform and to seek help—not only looking for surgery.

Many interesting aspects rise from the obstetric circumstances, forensic issues of all types of nerve injury, handling of children growing with a functionally impaired upper limb, socio-professional (re-)integration of adult patients, and development of interdisciplinary strategies, involving physio- and ergotherapists.

One might forget the moment when a carrier really starts and when the patient flow becomes regular without working actively on recruitment—but now, I see more clearly the last straight line unto the "end," the moment when the next generation will take over the challenge and job. Experience has grown over time, and now has come the time to communicate, to discuss, to summarize perhaps, or to build new strategies and tools. It took me years to elaborate my PhD work on the rotational imbalance of the glenohumeral joint following obstetric brachial plexus palsy, and it was only after more than 15 years of surgery dedicated to that complex issue did I imagine to cooperate with an industrial company to "create" a specific orthosis, individually adapted, to address the frequent posture of medial rotation contracture of the glenohumeral joint, associated or not with a supination contracture of the forearm. Once the requirements were clearly outlined and discussed with the manufacturer, the prototypes appeared rapidly, and now, we are refining details in a population of some 20 children—where this new tool offered and offers a change of paradigm and treatment strategy, as we may act in prevention when diagnosing the deformity early and counteracting the disbalance before any surgical indication. This event, like others, was not prepared, but just happened, because at a given right moment, both the idea and the potential builder met and agreed—a moment of great luck and joy of satisfaction.

11.3 Digital Medicine

As a student in mathematics and informatics in the mid-1970s, I participated in the shift from big computers to the raising personal computers—and their progressive way into our daily living. Now it is the time of digital patient record and hospital informatic department holding and transferring all relevant patient data, the administrative part, imagery data, and relevant protocols. The major argument is that modern medicine will improve with better data management, while patient-related data continue to increase.

Little concern is given to the benefit/cost balance (the huge hardware park, the new professionals required) and the risk of dependency, also concerning hackers and new criminals. Reports about blocked computer systems in big hospitals increase, and one single person may ask why this evolution is urged and constrained in exactly this direction, although we may easily imagine how difficult our work may be the day our computers fail—just because there is no electricity, a major hacker attack, or any bad coalition of perhaps tiny mistakes.

The discussion with the younger generation is not easy, as these colleagues grew up with the screen (and easily would understand our arguments as coming from old, inadaptable, inflexible people). All political responsibility is in line with this modern and unavoidable evolution into even more digitalization, especially in medicine, and the dependent economic aspect is even more in favor to collect all data and organize the medical "factory" in this computerized way.

There is no doubt that personal computers offer serious help and improvement, especially whenever we choose to seek their help. But we must stay aware of the risks and limitations and have realistic alternatives. Who drives his car to a new destination backing up the car GPS with an old-fashioned map? (I recently wondered when asking in a gas station to buy a county map that it is no longer available.) Whenever our cell phone has a hardware problem (the battery may not be recharged), we immediately experience a serious communication problem.

Will medicine improve by collecting and analyzing more and more data? Who came up with this argument (perhaps the computer industry herself)? Medicine is a relational job, and already now the young colleagues look more time onto their screen than into the patient's face—we have to imagine that this will seriously affect the quality of relation, dialogue, built confidence, and context for decisions. Medicine is a human science, full of typically human behavior. Empathy, concentration, and double rational and emotional binding … digitalization will not fight these behaviors but may become a serious burden, constraint, and source of alienation. I am not looking for arguments to stop or reverse the regular use of computers—but we need a serious information, discussion, and "road map" for a medical practice where the computer is a helpful tool and not a supermaster.

Whenever medicine becomes very individual, standardization has its limits and the place of digitalization too. Skilled surgery, and reconstructive surgery with microsurgical steps, remains an interesting lab for this evolution. Robot-assisted microsurgery is challenging routine human precision and reliability. Telemedical application of this principle will of course allow the dream that the expert is sitting in an office in a city in one country, while the robot and the operated patient are miles away, in another city hospital of another country. But is this really the way our evolution should move forward—or is there a need for debate and new paradigms?

My feeling over the last 20 years is that hospital medicine gets more and more economy driven and that digitalization of patient-related activities not only serves this aspect but more and more limits our liberty to live our job—the art of healing ("Heilkunst," "art de guérir"). There is a generation shift—as always—but is this beneficial for the patient and his demand, the care for his health? I believe that we will need to frame this evolution and reconsider the essentials of the unique interaction between a doctor and his patient: confidence, competence, and sincerity. Patient data are not the patient; they are an ever-growing adjunct we need to consider. But there are limits, and every caregiver knows that: there is a time needed to get the diagnosis and context, to read the reports, and to add analyses, imagery, and external advices. But once the story is complete, the diagnosis is certain, and the treatment plan is accorded, there is no more need for a continuously streaming data input. Our skill remains to do good data selection, extraction, and synthesis, building a frame, always rechecking it, and using informatics on demand, but not getting driven by a digital monster.

All arguments I read and heard so far for digitalization are presented in a brilliant intellectual rational way. It is difficult to formulate objections and precise limitations, to counteract with the same rationality. Perhaps, we should listen much more to general practitioners, all helpers in the first line—how medicine can succeed if the patient doubts or refuses the treatment, just because there is misunderstanding, lack of trust, and fear. Digitalization will not overcome these human barriers. Digitalization will never imagine all facets of a human being and its reactions.

Therefore, we should not fight against this technical revolution but should put limits and clear indications, while still being the masters of our art.

11.4 Managing Expectations Between Children with a Brachial Plexus Injury, Their Parents, and Medical Experts from the Experiences of a Support Group

The author is the mother of a son with an obstetrical brachial plexus injury. She is the chairperson of Plexuskinder e.V., a German support group that provides information, advice, and support to affected children and adults and their families. Speaking from more than a decade of experience, this section focuses on the relationships with medical experts, the affected children, and their parents.

Managing expectations between children with a brachial plexus injury, their parents, and medical experts from the experiences of a support group:

After a family is confronted with the outcome of a traumatic birth injury, a new relationship between them and the surgical specialist is formed. This might be a brief one-time encounter, but it might also become long, sometime lifelong ongoing relationship. As with all relationships, it is important to understand what each side is experiencing and what the relevant expectations are. The medical side might not be aware of the many questions, pressures, and challenges the families face. The affected persons and their families might have unrealistic expectations about what will be medically possible. Managing expectations of the medical and surgical treatment options is very important. There are many different challenges that affected children and their families face; here is a brief outline of a few:

There is a surgical option.

In many cases, parents are confronted with a previously fully unknown injury or illness. The fact that there might be a surgical option could be totally unknown to them. In many cases, they are not properly informed and educated. A surgical option might be surprising, confusing, and even menacing. The realization that the nonsurgical alternatives are not a solution might be rebutted by other medical experts. Parents frequently here things like the following: "It might get better on its own." "We will treat the injury away, with the right training it will solve itself." "Just sit tight and wait."

The family is surrounded by nonspecialists.

Once a surgical option is on the table, many families report the negative impact of questions and concerns of friends and family and even other medical professionals, who weigh in without detailed medical knowledge, further unsettling and confusing the parents. "Do you want to do that to your child?" "Surgery? But she is so small. You know, she might die on the table?"

Regret after a delayed surgical approach.

The information that there is a surgical option might come much too late, leading to further problems and possible disappointments. "We didn't know." "Nobody told us." "If we had known that there was a surgery it could have been all better now, it's all our/the doctor's fault."

Expectations from the surgery.

The expectations from the surgical treatment might be unrealistically high. "My child had the surgery you recommended, but the outcome is not satisfying. We wanted everything to go back to normal, as if nothing happened." "We are disappointed. When will everything be ok? Why is this taking so long?" "Can you make my child's arm look normal?"

Surgery is only one piece of the puzzle.

Parents might underestimate all the other factors that lead to a positive outcome after a successful surgery: regular therapy, increased awareness, attention to increase the arm's integration into everyday movements, sports, general health, attention to general physical fitness, psychological aspects, parenting skills, feeling of self-worth, self-confidence, and other relationships. Surgery is just one piece of the puzzle on the path to the highest possible quality of life with the injury. It is not possible to delegate therapy to a specialist once a week and expect the highest possible outcome. Therapy is play, and play is therapy and has to be integrated into daily life.

Emotional roller coaster.

What are the effects of the injury on the affected child, the parent, the marriage, and the siblings? How is the family dealing with them? Are they seeking out professional help? "Mom, stop crying, it's my arm."

Unexpected side effects.

Side effects might come as a surprise, both in the long term and in the short term. Postural adaptations and overuse of the unaffected side might cause problems many years later. Pain management could become an issue and should be addressed. "Why is my children having problems/pain in areas unrelated to the arm?" "The other arm hurts from using it too much."

Goal setting with attention to what the child wants.

After choosing the right surgery for the child, the following ongoing therapy has to be continuously adapted to the child goals both with a short-term and a long-term focus in mind. Goals should be focused on functional aspects—riding a bike, swimming, tying shoelaces, and getting dressed quickly—and not on outcomes totally abstract to a childlike degree of supination. As soon as the child is old enough to voice an opinion, this should be integrated in the therapeutic plan, since the parents might have quite different goals in mind. "What do you want to be able to do? What are your goals? What do you want to achieve? What for? By when? Why is this important to you?"

Regret, guilt, blaming.

Perhaps the most difficult issues parents face are understanding and accepting the situation and dealing with regret, guilt, blaming, and maybe even a legal battle. The fact that this has happened to their child is difficult to accept. It is vital that the medical experts understand the emotional distress that the injury and its consequences have caused and are causing. Professional help in dealing with the injury and its consequences should be encouraged for both the child and the parents.

Finding support in others.

Regular exchanges with other affected children and families, sharing experiences, speaking about worries, swapping ideas for little ways to make daily life better, and supporting and encouraging each other are of the highest importance. A support group serves as the collector of wisdom for the families and as the carrier of information towards the medical community. Regular exchanges between both groups help to work towards the common goal: the highest possible quality of life for the affected children.

11.5 Patient-Surgeon Interaction in the Twenty-First Century

From paternalism to cooperation.

A global view, including life philosophy, psychology, manual work, and overall patient wishes … the challenge of the meeting between two responsible adults;-))).

For centuries, doctors were godlike authorities, and they appreciated the "superior level" allowing them to decide, apply a treatment, or perform a surgery on a patient-object. The doctor knew what is good for the patient, decided which diagnosis to share (or not, like malignancy), and was the never-challenged decider.

Obstetricians knew how the pregnant woman should deliver and how to protect the baby. When medicine failed, it was destiny, and no discussion about the limits of skill came up.

Those times are definitely gone; the adult patient is our alter ego, a partner, and legal regulations control this equity.

We have to inform the patient in a way that enables himself to understand and decide among all alternatives, and even to take decisions on which we disagree—as he is the only manager of his own life.

This sounds logical, human, and evaluated—but is not the rule in all aspects of medicine or all countries. In reconstructive surgery, where the goal is never to escape from a life-threatening situation, it is even more important that the patient feels totally free to decide—even if at the end, he resigns from the proposed treatment plan. Issues are just patience (a time gap for further reflection), a second or third opinion, and the withdrawal of the elaborated plan.

Nowadays, patients are well documented and surf the internet (and doctor google) but are still willing to get convinced by sound counselling.

References

1. Bahm J, editor. Movement disorders of the upper extremities in children. Conservative and operative therapy. Cham: Springer; 2021.
2. Vulpius O, Stoffel A. Orthopädische therapie. In: Lewandowsky M, editor. Handbuch der Neurologie. Cham: Springer; 1910. p. 1299–321.
3. Kennedy R. Suture of the brachial plexus in birth paralysis of the upper extremity. BMJ. 1903;1(2197):298–301.

12 Conclusions

J. Bahm, *Surgical Rationales in Functional Reconstructive Surgery of the Upper Extremity*,
https://doi.org/10.1007/978-3-031-32005-7_12

35 Years ago, as a medical student thinking/deciding about a specialty, while discovering the different disciplines through the first practical clinical work, I hesitated between psychiatry and surgery—the former because of the concern beyond somatic medicine, and the latter because of the manual work and the great variety within surgical subspecialties.

Then, as a young trainee in surgery, I looked for hand and reconstructive surgery and I dedicated myself to children, because children always add unpredictable moments and fun where adults might get too serious and boring.

I actually tried to summarize the thoughts and knowledge for all those who are interested in the same way. I never intended to be or have been a basic searcher or gifted scientist. I searched to translate medical knowledge for common people, the patients, and their relatives, to make our diagnosis and treatment comprehensive.

This book tries to make my thoughts and strategy more comprehensive, allowing the newcomer to collect experience without redoing all my work and the mature colleague to oppose my ideas to his own experience and conviction. I put this on the table and do hope that there will be discussion, debate, birth of new ideas, or confirmation of strategies.

Studying and practicing medicine open the mind for all human sciences and gifted me with a curiosity for all aspects of this wonderful world, especially nature. Thus, a lot of different ideas flew into those chapters. Lots are inspired by contacts with patients or colleagues.

The actual developments of modern medicine and the speed of development in informatics sustaining our health system will change the face and body of medical science and surgical practice in the forthcoming twenty-first century—but I do hope that some of my strategies or rationales will remain and be helpful to the reader.

Also, I do repeat how grateful and happy I am for those great colleagues who shared their special knowledge and beautiful figures and drawings to make this book even more living.

Bibliography

Bahm J. Upper limb motion disorders: diagnosis and treatment. Springer; 2021.

Birch R. Surgical disorders of the peripheral nerves. Springer Science 2011 (2nd edition).

Brand PW, Hollister A. Clinical mechanics of the hand. St. Louis, MO: Mosby Year Book; 1993.

Dellon AL. Pain surgery. Springer; 2019.

Dörner K. Der gute Arzt. Schattauer 2001.

Lundborg G. Nerve injury and repair. Philadelphia, PA: Elsevier Churchill Livingstone; 2004.

Mackinnon S. Nerve surgery. Thieme 2015.

Moseley L, Butler D, Beames T. The graded motor imagery handbook. Adelaide, SA: NOI Group; 2019.

Schwenzer T, Bahm J. Schulterdystokie und Plexusparese. Springer 2016.

Spicher C. Douleurs neuropathiques—Evaluation Clinique et reeducation sensitive. 4.éd ed. Montpellier: Sauramps Médical; 2020.

J. Bahm, *Surgical Rationales in Functional Reconstructive Surgery of the Upper Extremity*, https://doi.org/10.1007/978-3-031-32005-7

<u>GPSR Compliance</u>

The European Union's (EU) General Product Safety Regulation (GPSR) is a set of rules that requires consumer products to be safe and our obligations to ensure this.

If you have any concerns about our products, you can contact us on ProductSafety@springernature.com

In case Publisher is established outside the EU, the EU authorized representative is:

Springer Nature Customer Service Center GmbH
Europaplatz 3
69115 Heidelberg, Germany

Batch number: 10406427

Printed by Printforce, the Netherlands